Other titles in this seri

Island Ecology
Plant–Atmosphere Relations
Modelling
Vegetation Dynamics
Animal Population Dynamics

R. Moss, A. Watson and
J. Ollason

Outline

Editors

George M. Dunnet
Regius Professor of Natural History
University of Aberdeen

Charles H. Gimingham
Professor of Botany,
University of Aberdeen

Editors' Foreword

Both in its theoretical and applied aspects, ecology is developing rapidly. In part because it offers a relatively new and fresh approach to biological enquiry, but it also stems from the revolution in public attitudes towards the quality of the human environment and the conservation of nature. There are today more professional ecologists than ever before, and the number of students seeking courses in ecology remains high. In schools as well as universities the teaching of ecology is now widely accepted as an essential component of biological education, but it is only within the past quarter of a century that this has come about. In the same period, the journals devoted to publication of ecological research have expanded in number and size, and books on aspects of ecology appear in ever-increasing numbers.

These are indications of a healthy and vigorous condition, which is satisfactory not only in regard to the progress of biological science but also because of the vital importance of ecological understanding to the well-being of man. However, such rapid advances bring their problems. The subject develops so rapidly in scope, depth and relevance that textbooks, or parts of them, soon become out-of-date or inappropriate for particular courses. The very width of the front across which the ecological approach is being applied to biological and environmental questions introduces difficulties: every teacher handles his subject in a different way and no two courses are identical in content.

This diversity, though stimulating and profitable, has the effect that no single text-book is likely to satisfy fully the needs of the student attending a course in ecology. Very often extracts from a wide range of books must be consulted, and while this may do no harm it is time-consuming and expensive. The present series has been designed to offer quite a large number of relatively small booklets, each on a restricted topic of fundamental importance which is likely to constitute a self-contained component of more comprehensive courses. A selection can then be made, at reasonable cost, of texts appropriate to particular courses or the interests of the reader. Each is written by an acknowledged expert in the subject, and is intended to offer an up-to-date, concise summary which will be of value to those engaged in teaching, research or applied ecology as well as to students.

Studies in Ecology

Insect Herbivory

I.D. HODKINSON
and
M.K. HUGHES
Department of Biology
Liverpool Polytechnic

CHAPMAN AND HALL
LONDON NEW YORK

First published 1982
by Chapman and Hall Ltd
11 New Fetter Lane, London EC4P 4EE
Published in the USA
by Chapman and Hall
733 Third Avenue, New York NY 10017

© 1982 I.D. Hodkinson and M.K. Hughes

Printed in Great Britain by
J.W. Arrowsmith Ltd, Bristol

ISBN 0 412 23870 5

British Library Cataloguing in Publication Data

Hodkinson, I.D.
  Insect herbivory.—(Outline studies in ecology)
  1. Insects—Feeding and feeds
  I. Title   II. Hughes, M.K.   III. Series
  595.7′013   QL496

  ISBN 0-412-23870-5

Library of Congress Cataloging in Publication Data

Hodkinson, I.D. (Ian David)
  Insect herbivory.

  (Outline studies in ecology)
  Bibliography: p.
  Includes index.
  1. Insect-plant relationships. 2. Insects—Host plants.
3. Insects—Food.   I. Hughes, M.K.   II. Title.   III. Series.
QL496.H63   1982   595.7′053   82-9525
ISBN 0-412-23870-5   AACR2

Contents

Preface

This book attempts to summarize what we know about insect–plant relationships without becoming too involved with untestable hypotheses. It is not intended to be comprehensive and we have deliberately excluded discussion of aquatic organisms and fungi. Our definition of insect herbivores is intentionally broad. It includes all insects which feed on plants, although we have emphasized those which feed primarily on the photosynthetic tissues. Some reference is made to seed predation but pollination ecology is excluded.

We thank Drs P.H. Smith and M. Luxton for their helpful comments on the manuscript but we accept full responsibility for any mistakes which may remain. Finally, we thank the various publishers and authors who gave us permission to use copyright material.

1 Introduction

The net primary production of the 300 000 species of vascular plant which inhabit the dry land surface of the earth has been estimated at about 115×10^9t per annum. This represents a massive resource potentially available for exploitation by the herbivorous insects, which themselves probably number in excess of 500 000 species.

The impact of insects on agricultural crops has been documented since biblical times when plagues of locusts were 'all over the land of Egypt'. In contrast an understanding of their impact on and interaction with natural vegetation has only really developed over the last hundred years. In natural ecosystems widespread outbreaks of herbivorous insects, leading to complete defoliation of vegetation, happen only sporadically. The most spectacular examples tend to occur in the low diversity forests of the cool temperate or subarctic regions of the earth. For example, within the last century population eruptions of the moth species belonging to the genera *Oporinia* and *Operophtera* have lead to widespread defoliation and death of birch forest in northern Finland and along the Scandinavian mountain chain [1]. It has been suggested that such outbreaks are rare in the more diverse tropical forests but recent work in Panamanian lowland rain forest, involving the moth *Zunacetha annulata* and its larval host plant tree *Hybanthus prunifolius*, has shown that such outbreaks do occur, but they are less visually obvious in such habitats [2]. Thus while heavy defoliation of plants does occur it is the exception rather than the rule: in general, phytophagous insects consume only a fraction of the available primary production.

This book sets out, therefore, to examine the ways in which herbivorous insects exploit their food resource and the means by which plants seek to minimize their depredations. It attempts to present a quantitative analysis of insect herbivory set against the background of the ecological community and ecosystem.

1.1 The evolutionary perspective and its implications

The earliest known forms of insect are thought to have been detritus feeders. The habit of plant feeding appears to have evolved independently on several occasions, even within a single order such as the Hymenoptera [3]. In consequence the herbivorous habit is widespread, but disjunctively displayed, across the different insect groups. Fig. 1.1 illustrates the evolution of the main plant-feeding insect orders relative to that of the major plant groups. Particular attention should be paid to the fact that the flowering plants or Angiosperms, the group which

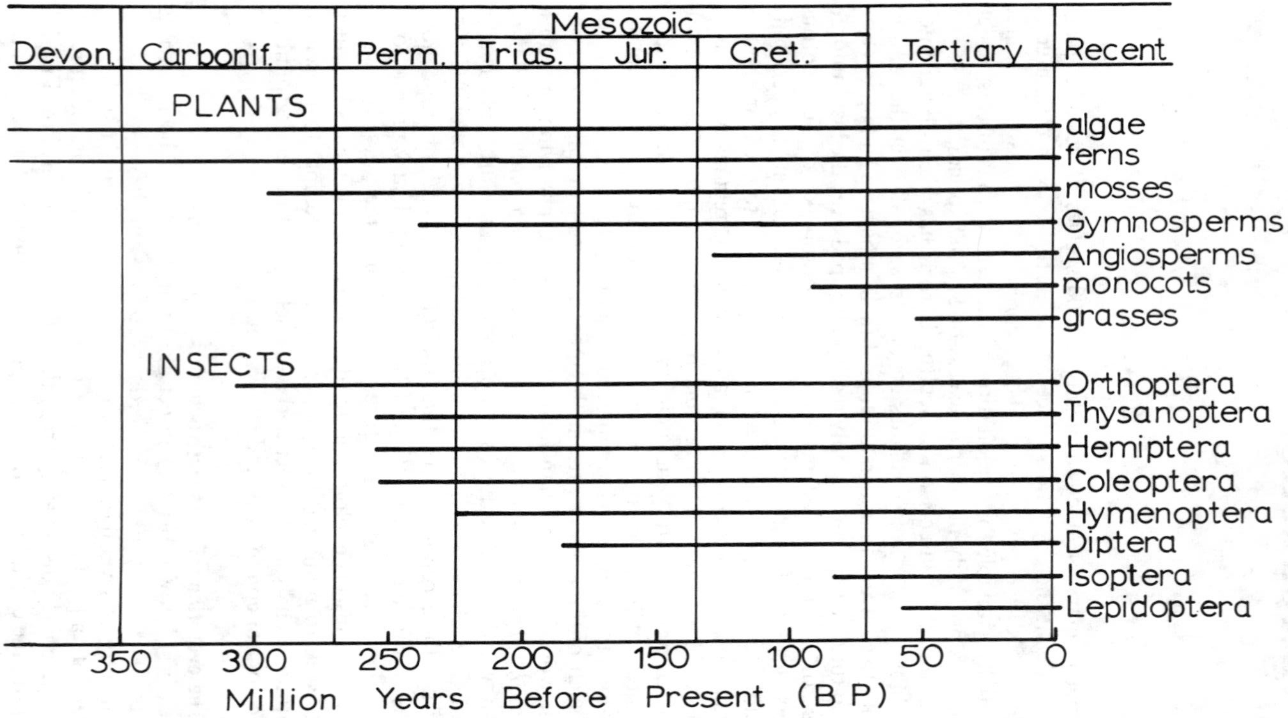

Fig. 1.1 Evolutionary chronology of the major orders of herbivorous insects and the major groups of living plants.

dominates most contemporary floras, did not evolve until the early Cretaceous, about 125 million years before the present (BP). During the Cretaceous the Angiosperms underwent explosive evolution, largely displacing the pre-existing flora over most of the globe and this provided a major impetus for the evolution of the phytophagous insects. Some insect orders such as the Lepidoptera (butterflies and moths) and the Isoptera (termites) are thought to have evolved subsequent to the Angiosperm explosion [4]. Other orders with an older fossil record, such as the Hemiptera (bugs) and Diptera (flies), appear to have contained evolutionary stocks which transferred onto the angiosperms during the Cretaceous and evolved rapidly to give rise to several important extant phytophagous families, such as the aphids (Aphididae) and the leaf-mining flies, the Agromyzidae.

The insects have thus undergone a long and varied period of coevolution and coadaptation with their host-plants and it is not surprising that different groups of insects have developed different patterns of host-plant associations coupled with the different life cycle strategies and feeding mechanisms necessary for the exploitation of their hosts. Even in closely related groups such as the aphids (Aphidoidea) and the jumping plant lice (Psylloidea) the distribution of host relationships across the plant kingdom often differs. The psyllids are restricted almost exclusively to the Dicotyledons whereas the aphids in addition occur commonly on both the Monocotyledons and the Coniferae.

Table 1.1 shows the host specificity of a number of different insect groups chosen to represent a variety of feeding mechanisms, including leaf chewing, sap sucking and seed/fruit feeding forms. The monophagous category has, of necessity, been broadly defined to include species which feed on host plants in a single genus. The oligophagous category includes species restricted to hosts within a single plant family whereas species feeding on plants in more than one family are termed polyphagous. A host-plant is here defined as one on which the insect completes its

Table 1.1 Host-plant range of selected insect herbivore groups. (Sources [5, 6, 7, 8, 9, 10, 11, 12])

Insect group	Sample size (species)	Phagism category (%)		
		Mono	Oligo	Poly
Nearctic butterflies	244	48	20	32
British thrips	120	47	22	31
British aphids	528	76	18	6
British leafhoppers on trees	55	73	4	23
British psyllids	77	79	21	0
Costa Rican seed/fruit beetles	85	89	11	0
British sawflies	396	65	23	12
Costa Rican forest grasshoppers	26	38	27	35

growth and development. In some insects such as the Lepidoptera the larvae may feed on very few plants but the adult may take nectar from a wide variety. Most species of herbivorous insect exhibit a close association with a fairly narrow range of host-plants; that is they are monophagous or oligophagous as defined above. Broad polyphagy appears to be less common although in some individual groups such as the Orthoptera it is perhaps more predominant [13]. Occasionally, within a predominantly monophagous group, a species may exhibit very broad host preferences; a good example is the peach–potato aphid *Myzus persicae*, which has been recorded from over 50 different plant families. Often groups of closely related host-specific insect species will feed on groups of closely related plant species, indicating a close evolutionary relationship between the two groups. For example, the different subgenera of the leaf-beetle genus *Chrysolina* are usually restricted to just one or two families of host-plant [16]. However, this is not always the case and in some more recent groups such as the Lepidoptera the evidence suggests that host-plant switching onto distantly related plant groups has frequently occurred [14, 15].

Any plant species will, through evolutionary time, have 'gathered' its own specific insect fauna, with the different species exploiting the plant

Table 1.2 Feeding site and damage symptoms of insects on *Epilobium angustifolium*. (After Myerscough 1980, with additions)

Insect		Feeding site
Leaf chewers		
LEPIDOPTERA		
Sphingidae	*Deilephila* (2 spp.)	Leaves growing
Momphidae	*Mompha* (2 spp.)	shoots or leaf surface
COLEOPTERA		
• Chrysomelidae	*Altica* (2 spp.)	Leaves and growing shoot
Sap suckers		
HEMIPTERA		
Aphididae	*Aphis* (5 spp.)	Phloem of leaf or stem
	Macrosiphum euphorbiae	
Aphalaridae	*Craspedolepta nebulosa*	Phloem of leaf or stem
	Craspedolepta subpunctata	Phloem of root causing galls
Miridae	*Lygocoris pabulinus*	Mesophyll of leaf
Cercopidae	*Philaenus spumarius*	Xylem of stem
Other gall formers		
LEPIDOPTERA		
Momphidae	*Mompha nodicolella*	Stem gall
DIPTERA		
Cecidomyiidae	*Dasineura* (2 spp.)	Flower bud or leaf margin gall

in a variety of ways. Table 1.2 shows the characteristic feeding sites of the insects associated with the common herbaceous perennial plant *Epilobium angustifolium* (rose-bay willow herb) in Britain [17]. In some tree species such as oaks (*Quercus* spp.) the number of associated insects may be as high as three hundred species [18].

2   Plants as food for insects

Plant tissues are largely made up of water and relatively indigestible compounds such as cellulose and lignin. This makes them a good potential source of water, but in many cases an unpromising source of energy and nutrients. Nevertheless, plant tissues are present in great abundance and variety and they form the food resource for many thousands of insect species. What sort of foods do plant tissues offer to insects?

The fresh weight of leaves tends to be made up of more than 90% water and 1 to 3% protein: most of the residue is carbohydrate. Seeds and especially pollen have a lower water content and more protein, whereas other tissues such as wood contain less protein. Almost all higher plant tissues contain markedly lower concentrations of protein (< 10% fresh wt.) than insect tissue (5–20% fresh wt.) [19]. These and other differences in the gross biochemical composition of the respective tissues are reflected in their energy content (Fig. 2.1). Insect tissues with their higher fat and protein contents show a range of energy contents (Joulerific values) exceeding those of almost all plant tissues. Typically, they contain about 22 to 28 kJ g^{-1} dry wt., whereas deciduous tree leaves contain 17 to 22 kJ g^{-1} dry wt. Most leaf material has an even lower energy content (16 to 18 kJ g^{-1} dry wt. [20, 21, 22]. Similar comments apply to the supply of the elements characteristic of protein. Few plant tissues contain levels of nitrogen or phosphorus comparable with those in insect tissue (Fig. 2.2). Most plant materials contain at most 3 to 4% nitrogen on a dry weight basis while insect tissue contains 7 to 14% [24].

In terms of its gross composition, most plant tissue is fairly low grade food for insects. Thus to acquire the quantities of energy, nitrogen and often phosphorus they need, insect herbivores must consume disproportionately large quantities of plant for each unit of insect growth. As well as being forced into this high consumption strategy they must also utilize foods which meet their requirements for certain specific organic compounds and trace elements. In most insects the amino acid methionine and all the water soluble B-vitamins are essential. The quantities of these compounds to be found in plant tissues vary. All insects need small quantities of sterols and they have to be able to convert plant sterols to those they need [19]. Furthermore, certain metals, such as copper and zinc, are essential in trace quantities. Their concentration in tissues may vary enormously (Fig. 2.2) and it is known that some invertebrates, such as isopods, are able partially to regulate their body concentration of some of these elements by modifying their

15

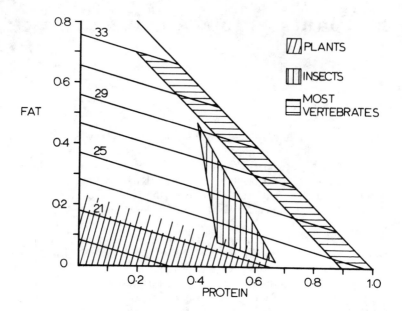

Fig. 2.1 The relationship between the fat and protein content of various groups of organisms (expressed as proportions of unit dry weight along ordinate and abscissa) and their energy content in kJ g^{-1} dry weight indicated by the sloping lines. The half-hatching extending above 21 kJ g^{-1} for plants indicates that leaf tissues may exceed this value for short periods. (Modified from Southwood [19], reproduced by permission of the Royal Entomological Society of London.)

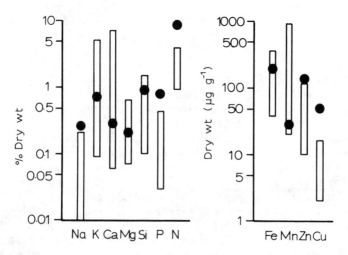

Fig. 2.2 Concentrations of elements in the green tissues of seed plants (vertical bars show ranges) and typical insects (closed circles). Note log scales. (Based on data from Allen *et al.* [23].)

rates of feeding [25]. It may well be that the requirement for trace metals also affects insect feeding biology.

2.1 Variation in the nutritive value of plant tissue

Some plant tissues offer a better food source than others in terms of energy and nitrogen content. As already mentioned seeds and pollen have a relatively high protein content which may be expressed as a high energy and/or nitrogen level. Active meristems, being made up of young rapidly dividing cells are also energy and nitrogen-rich, although for reasons discussed later (Chapter 5), relatively few insects use them. Rapidly growing tissues, such as new shoots, roots and young leaves, resemble meristems in their energy and nitrogen levels. In many plants the proportion of crude protein in leaves decreases as the leaves age, while that of structural carbohydrates increases and fats remain more or less constant [26]. Sucrose content may rise as a result of all this. The energy content of, for example, birch leaves may fall from 21 kJ g^{-1} dry wt. when newly opened to 18.8 kJ g^{-1} dry wt. at the end of the summer [20]. Most of this change takes place in the first six weeks of leaf life. The fall in total nitrogen may be much greater. In northern tree-line birch, total leaf nitrogen fell from 3% dry wt. in June to under 1% dry wt. in September [27]. Similar trends occur in non-woody plants. For example, new shoots of the grass *Holcus mollis*, whether formed in spring or autumn, have elevated nitrogen levels [28, 29] when compared with existing shoots. While plant tissues in general may contain 3% dry wt. nitrogen or less, young fast growing tissues may contain up to 7% dry wt. These high total nitrogen levels occur for short periods of time, for example, the first few weeks of a growth season. Plant tissue in this relatively nitrogen and energy rich condition presents a distinct food advantage to any insect herbivore as compared with other plant material.

Clearly, total energy content and total nitrogen are fairly crude indicators of the potential food resource available to a consumer. At times of rapid tissue formation plants must mobilize materials from various organs and transport them to the point of growth. Consequently, *soluble* nitrogen levels will be particularly high at times of rapid protein synthesis, such as leaf growth and at times of leaf senescence when materials are withdrawn into the stem [29]. Such changes also occur in the soluble nitrogen and amino acid levels of the evergreen conifer sitka spruce [30]. High levels of soluble nitrogen may well render a plant tissue an advantageous food to an insect, since it will provide easily absorbed amino acids. The flux of amino acids and sugars from their source to the point of use provides an important and specialized food resource for some sap-sucking insects. These insects, which tap the phloem sap, do not directly damage or reduce the plant's productive machinery but they sequester some of the photosynthetic product. In certain circumstances this may increase the rate of production at the source and thus this feeding strategy may have considerable advantages. Changes in the composition of phloem sap can be considerable. For example, the nitrogen content may vary from 0.004 to 0.6% N w/v [24] whilst sucrose may vary from 1.7 to 8.6% w/v [31]. Much of this variation is associated with the difference between quiescent periods and periods of active growth.

The positive nutritional attributes of plant tissues such as their energy or nitrogen content may vary in other ways. Plant nitrogen may decline with plant age and vary with plant density [24]. Considerable differences exist between the energy and nitrogen contents of analogous organs in different plant species [20, 32]. It has been suggested that if carnivorous plants and those with nitrogen-fixing symbionts are excluded, plants on impoverished or exposed substrates are stress-selected. One attribute associated with such plants is their ability to survive on little nitrogen [24]. This phenomenon, however, must be distinguished from changes in plant nitrogen content associated with environmental stress, such as drought, acting within the plant's lifetime (see Chapter 7).

2.2 Barriers to the use of plant tissues

The mere fact that a plant tissue contains the right quantities of energy and nutritive materials in a suitable mixture does not mean that it is either a suitable or an available food for a particular insect species. A variety of barriers exist between the insect and the resources of energy and nutrients contained in the plant. Members of the insect population must be able to find their food reliably. Since plant food is often of low quality, insect herbivores cannot afford the luxury of searching and hunting for elusive and unpredictable supplies. A plant tissue may be easily locatable or 'apparent' to an insect species in a number of ways. It may be large, exist at high density, or with great regularity. It may be abundant, occur frequently in time, or with great temporal regularity. In a temperate forest the leaves of the dominant tree species are 'apparent' to insect herbivores on most of these counts, although young leaves are less apparent than older leaves. In contrast, ruderal plants, those invading recently disturbed land, are singularly 'non-apparent' to insect species populations. Some attributes of apparent and non-apparent plants are summarized in Table 2.1.

Table 2.1 Characteristic apparent and non-apparent plant materials

Apparent	Non-apparent
Woody perennial	Annual species
Climax species	Pioneer species
Common species	Rare species
Mature leaves	New leaves
Bark, stems	Leaves
Evergreen leaves	Deciduous leaves

A plant tissue may be placed nearer to the non-apparent end of this spectrum by virtue of individual variation in phenology. For example, many leaf-chewing Lepidoptera larvae living on oak in Europe and the eastern USA must be present and consuming rapidly in the first few weeks after bud burst if they are to gain an adequate nitrogen supply for their development (see below). Larvae hatching too early find inadequate food supplies and risk high mortality from factors such as frost. Those hatching a few days too late may not reach the later instars before the quality of the leaf tissue starts to decline rapidly and dramatically. The risks to the insect population are greatly increased by the high degree of variability in the timing of bud-burst that may exist between individual oaks in the same forest [33, 34]. At a site in northern England, the time of bud-burst of twenty-five mature oaks varied by up to four weeks in one year.

Some plant organs may possess structures which repel, injure or kill insects landing on them. Bean cultivars vary in the density of hooked hairs (trichomes) on their surface. Leafhopper nymphs may become impaled on these hooked hairs and their survival is lowest on the cultivar with most hooked hairs [36]. In other species, glandular hairs produce sticky secretions which impede insect movement, or secrete substances toxic to the herbivore [35]. Plant species also vary greatly in the mechanical properties of their leaves. Leaves with a heavy 'toothed' cuticle, such as holly (*Ilex aquifolium*) are more difficult to chew than those with a light cuticle such as hazel (*Corylus avellana*). These mechanical properties may vary within the growth season as a plant organ ages, and between leaves in different parts of the canopy [34].

A plant may exist in a mutualistic association with another organism which repels consumers and thereby reduces the levels of herbivory. The best known example is the relationship between certain tropical *Acacia* species and the thorn acacia ants of the genus *Pseudomyrmex*. The plant shelters the ants in special hollow thorns and supplies food from extrafloral nectaries and specialized protein sources, the beltian bodies. In return the ants protect the plant against attack by herbivorous insects and competition from other plants [37, 38]. This phenomenon is not restricted to the tropics; ants are attracted to extrafloral nectar secreted by the aspen sunflower (*Helianthella quinquenervis*) growing in the temperate USA. The ants effectively disrupt oviposition by some of the insect seed predators of this plant, particularly at higher altitudes [39].

2.3 Trace compound barriers

The most intensively studied barriers between insects and potential plant food are the secondary metabolites. These are substances present in plant tissues which do not appear to play a central or major role in the basic physiological processes of the plant. Many have distinctive odours or tastes, or are coloured and so their presence may be easily detected. A considerable number are known to be toxic to insects, bringing about injury or death, depending on the circumstances and the quantity consumed. Amongst the best known compounds are the alkaloids. These are heterocyclic nitrogen compounds that exist as water-soluble cations [40]. Nicotine, cocaine, quinine, morphine and caffeine are all well-known alkaloids. Several plant families, mainly dicotyledons, contain alkaloids but the majority (approximately 70%) do not. Alkaloids occur more frequently in herbs than trees and more frequently in annuals than perennials. The concentration and nature of the alkaloids may vary between tissues, between individuals and between breeding populations within a species [41]. In general, alkaloid content is highest in enlarging and vacuolating cells and lowest in senescent cells. There are many other sources of variation in alkaloid content associated with the physiological state, stage of development and environmental conditions of the plant. Such variability has been studied in the lupin (*Lupinus*). Heavy infestations of thrips (Thysanoptera) were found on plants lacking alkaloids, whereas plants containing alkaloids were without thrips [42]. This phenomenon may result from a repellant effect due to the alkaloids, a toxic effect, or some combination of the two.

Repellant effects may be very specific. For example, α-tomatine is the characteristic alkaloid of the tomato plant (*Lycopersicon*). It repels the Colorado beetle (*Leptinotarsa decemlineata*) and the potato leafhopper (*Empoasca fabae*) but not the beetle *Epilachna* or the grasshopper *Melanoplus bivittatus*, even though the alkaloid is highly toxic [40, 43]. The Colorado beetle detects the presence of tomatine very much more efficiently than solanine, an alkaloid to which it is specifically adapted. Thus, tomatine acts as an effective 'no-entry' warning sign to this beetle. Similarly, Colorado beetles do not appear to attack the domestic potato's wild relative *Solanum demissum* [44]. This is due to the repellant action of alkaloids known as leptines rather than to the toxic action of the species characteristic alkaloid demissin [45].

Alkaloids need not be present in great quantity to have such effects. Along with other groups of secondary metabolites such as glucosino-lates and cyanogenic compounds, they tend to be present at less than 2% dry wt. [46]. The glucosinolates are organic nitrogen compounds which exist as anions and are present in all members of the plant family Cruciferae, which contains the Brassicas, and some species from other families. Glucosinolate concentrations in crucifers range up to 0.1% dry wt. They are found in association with the thioglucosidase enzymes, although separated from them structurally. These enzymes hydrolyse glucosinolates to a variety of products, always including D-glucose and HSO_4^-. The other products vary with a range of conditions and include isothiocyanate, nitrite or thiocyanate. When a herbivore chews a crucifer it brings the substrate and enzyme into contact in an aqueous medium and some mixture of these products results. The particular mixture of olfactory signal and toxicity characterizes the plant and determines the insect's response [47]. Interestingly, cultivated crucifers such as the domestic cabbage contain far lower concentrations of toxic glucosinolates than wild cabbage (*Brassica oleracea*) and are much more susceptible to insect attack [48].

20

When the tissue of other plant species, such as *Sorghum* and *Lotus corniculatus*, are crushed, highly toxic prussic acid (HCN) may be released. This results from the action of enzymes on trace quantities of carbohydrate derivatives known as cyanogenic glycosides [49]. There are a number of other groups of compounds found in plants which may have similar repellent or injurious effects on insect herbivores, even though present at only trace levels. These include chemical analogues of insect hormones which may disrupt the insect life cycle. The woody parts of the fir *Abies balsamea* produce an analogue of the juvenile hormone active in the plant bug *Pyrrhocoris apterus* [50]. The phytoecdysones, the plant analogues of the insect moulting hormone, are also found in a wide range of plants, such as *Podocarpus*. These plants tend to be perennial and woody, rather than annual and herbaceous [50].

There is such a wide range of repellent or disruptive trace substances present in plants that they must constitute, by their biological potency and diversity, a formidable range of barriers to feeding. Over a period of 270 million years of coevolution a wide range of interactions between insects and the plants possessing these compounds have developed.

2.4 Dosage-dependent chemical barriers

The trace compounds discussed in Section 2.3 need only be present in minute concentrations to influence food choice by insect herbivores. Other compounds such as tannins, resins and silica may act in a rather different dosage-dependent way, such that the degree of herbivory is directly related to their concentration in plant tissues. Although these substances may be active at concentrations as low as 1% fresh wt. they may be present at much higher levels [51, 52, 53]. Tannins are phenolic compounds found in all vascular plant groups. They bind to soluble proteins with the effect that they reduce enzyme activity and the availability of protein substrates to enzymes. This may directly affect the availability of the protein to insect herbivores [54]. The concentration of tannins in oak leaves increases from just under 1% dry wt. in April to near 2% in August and over 5% in September [34]. As little as 1% oak leaf tannin in an artificial diet produced a significant reduction in larval growth rate and pupal weight of the winter moth *Operophtera brumata* which normally feeds on oak leaves [51]. That this is a result of complexing of tannin with insect digestive enzymes and plant protein is supported by an examination of the seasonal trend of the protein:tannin ratio in oak leaves. In April, May and early June when the activity of leaf-chewing insects is greatest, the ratio exceeds 10:1. Thereafter, when herbivory is minimal, the ratio declines from 10:1 to 2:1 by early autumn [34]. It appears that the combination of declining protein content and decreasing protein:tannin ratio results in a marked reduction in the availability of oak leaf protein to winter moth caterpillars after the first few weeks of leaf life. This effectively narrows the time in which the insects can acquire an adequate nitrogen supply. However, it is important to exercise caution in interpreting the role of tannins in plants [55]. When different grasshopper species were fed wheat leaves with or

21

without added tannin, no effect on digestion was found [56]. Only in the case of hydrolysable tannin and the grass-eating *Locusta migratoria* was any toxicity established. This species is not normally exposed to significant tannin concentrations and it can be argued that as the grasshoppers evolved early in insect history, in a period where condensed tannins were present in all primitive vascular plants, they must have acquired the ability to tolerate these compounds. Furthermore, although trees of the genus *Eucalyptus* contain high levels of tannins and other phenols they are subject to considerable herbivory. Experiments on the larvae of the chrysomelid beetle *Paropsis atomaria*, showed that tannin or phenol concentrations had no effect on feeding rates or nitrogen use efficiencies [57]. This may be related to the evolutionary age of the Coleoptera, to the high pH in their gut, to the specific nature of *Eucalyptus* proteins or some detoxification mechanism in the beetle. Tannins also appear to stimulate feeding in the larvae of some Lepidoptera and it is possible that the role of tannins in plant defence has been overemphasized [55].

2.5 Changes in plant tissue resulting from insect herbivory

The possession of chemical barriers by a plant can be interpreted as a defence mechanism against insect feeding which the plant has acquired by natural selection. It is difficult, however, to test this assertion. Secondary plant metabolites have alternative functions ascribed to them by biologists in other fields. In particular, many are thought to be antimicrobial agents protecting plants from disease. This is especially true of lignin which renders xylem highly resistant to both animal digestion and microbial attack.

There are, however, examples where insect feeding has been shown to induce an increase in the concentration of some secondary metabolite in the remaining plant tissue, which then becomes more resistant to insect attack. Defoliation by the larch bud moth (*Zeiraphera*) results in a delayed leaf flush in *Larix decidua* and a change in leaves produced. They are smaller and tougher than those produced before defoliation and have lower nitrogen, higher fibre and resin content. This is, in turn, associated with higher larval mortality and lower fecundity of bud moth the next season [58, 59]. Damage to leaves of the downy birch, *Betula pubescens* results in an increase in the total phenolic content of neighbouring undamaged leaves within 2 days. This has been associated with delayed pupation in the moth *Oporinia autumnata*. Defoliation is followed next season by delayed leaf flush and smaller leaves which apparently results in later pupation and lower pupal weights in the moth [60]. These responses, which are induced by the feeding insect, seem to be analogous to the production of phytoalexins in plants invaded by pathogenic microbes.

2.6 Strategies of insect herbivory and plant response

The different kinds of barriers to herbivory discussed in this chapter are not independent of one another. In general, apparent tissues (those that insects may find reliably) contain dosage-dependent chemical barriers such as tannin. Some authors [52] view these as *quantitative plant defences*. These may be distinguished from the trace substance barriers such as the alkaloids or glucosinolates found typically in non-apparent plant tissues. These are seen as *qualitative plant defences*. It has already been pointed out that the presence and levels of these substances may be explained as a consequence of selection pressures other than herbivory. Bearing this in mind, a very brief summary will be given of current theory on plant responses to herbivory [46].

The general assumption is made that greatest evolutionary fitness for the plant follows from a minimization of herbivory. Resources taken by herbivores might otherwise be used for the production of a greater number of viable offspring. Thus, any modification in plant development, chemistry or ecological strategy that minimized net loss to herbivores will confer a selective advantage. Each defensive strategy, however, has its own cost to the plant. Energy or other resources which the plant diverts for defence cannot be used for growth and reproduction. This cost must be balanced against the benefit gained from the defence. Qualitative barriers should, therefore, have a lower metabolic cost to the plant than the quantitative barriers [56, 61]. Apparent plants living in stressful environments may adopt an induced response. Chemical barriers to herbivory and the associated metabolic cost to the plant, may remain low until insect damage occurs and then be raised to inhibit further attack. The response of birch to *Oporinia* damage (Section 2.5) illustrates an induced defence in an environment poor in resources.

Non-apparent plant tissues represent an unreliable food source and it can be argued that generalist (polyphagous) insects are best adapted for their exploitation since they can switch from one plant to another as available. In contrast apparent plant tissues represent a more predictable food source to which specialized (oligo- or monophagous) insects can become highly adapted. Plants in turn may evolve appropriate defences against these different herbivore strategies. The predicted response for non-apparent plants would be the development of low-cost, highly species – or tissue – specific defences such as the trace substances. Different plants would evolve their own specific defensive compounds and these defences can be seen as divergent. In contrast, the predicted response for apparent plants, which must defend themselves against a smaller number of specialist insects, would be the development of fairly general chemical defences. These are likely to be similar, or convergent, defences such as the dosage-dependent substances. Thus we reach a stage in the argument at which the defences of non-apparent and apparent plant tissues are divergent and convergent respectively. However, the plant defences in turn may evoke a further evolutionary response from the herbivores. Divergent defences would now begin to select for specialist herbivores which could overcome the specific

23

chemicals in particular plant species. Convergent defences would begin to select generalist insects which could overcome common quantitative defensive compounds such as tannins.

A diagrammatic representation of this proposed scheme of coevolution is shown in Fig. 2.3. The net outcome of these ideas is that the relative levels of pressure by specialist and generalist herbivores on apparent and non-apparent plant resources should be similar [46]. This hypothesis is based on very limited evidence and will be very difficult, if not impossible, to test.

The theories of plant defence do not take into account the possibility that herbivores may benefit plants. Trace compounds may attract as well as repel insects. A plant group may have coevolved with herbivores for most of its evolutionary development so that a degree of mutualism has been established [62]. Thus grasses are particularly well adapted to withstand high levels of herbivory, both in their growth form and physiology [63]. The existence of mutualism is well established in the case of flowers and insect pollinators. It could be that analogous parallel relationships have developed with insect herbivores. Some possible implications of this idea will be discussed in Chapter 7.

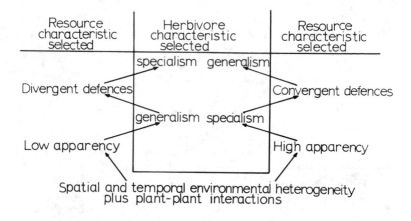

Fig. 2.3 Proposed scheme of coevolution between plant-defences and insect herbivores. For full explanation see text. (Reproduced from Rhoades [46] by permission of Academic Press.)

3 Insect adaptations to herbivory

Before feeding or oviposition any herbivorous insect is faced with the problem of locating its host plant both in space and time. It must be able to select its particular food plant(s) from amongst the complex array of available species and it must ensure that its period of feeding activity coincides with periods of plant availability. This is often further complicated by the need to select a particular plant organ or tissue at a specific phenological stage of development.

This chapter examines the behavioural mechanisms involved in host selection and investigates the problem of synchronization between insect and plant life cycles. It discusses the ways in which insects feed on plants and utilize the food obtained for growth and development.

3.1 Finding the food: host-plant location and recognition

Each plant species has its own particular odour, taste, colour and form and these are basic sensory cues which insects use to locate and recognize their hosts. Host-plant selection has been described as a catenary or chain process involving a number of successive steps which ultimately result in the insect feeding or ovipositing on the plant. Each step involves a behavioural response to specific stimuli: if the insect perceives the stimuli to be correct then it moves on to the next step but if the stimuli are wrong, the 'chain' is broken. Fig. 3.1 gives a generalized summary of the mechanisms of host selection involving four main steps, dispersal, attraction, arrest and feeding. In insect groups such as the Hemiptera in which both larvae and adults are herbivorous this may be followed by oviposition. For groups such as the Lepidoptera and Diptera, in which only the larvae are herbivorous, oviposition by adults usually follows the arrest stage.

The initial stage in host selection involves dispersal of the adult insect into the habitat in response to a variety of stimuli which may trigger a dispersal flight [64]. This merely serves to bring the insect within range of its host-plant. Attraction to a specific host-plant (step 2) involves either olfactory or visual stimuli or a combination of both. Attraction in some aphids appears to be visual and not specific: they respond to the yellow green colour of the vegetation [65]. Olfaction is more complex and involves the reception of volatile chemicals released from the plant, often at low concentrations. These chemicals stimulate the insect to fly upwind to the source of the odour, a so-called anemotaxis behavioural response. The chemical stimuli are perceived through sensilla located on the insects' antennae and mouthparts. The insect is prevented from

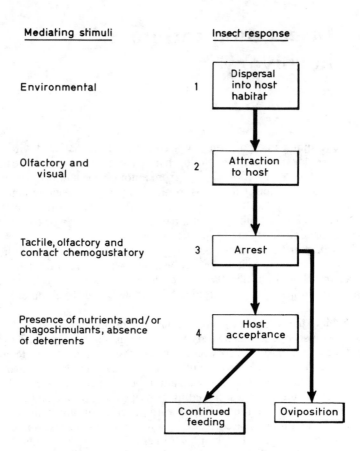

Fig. 3.1 Schematic representation of host-plant selection by herbivorous insects. For full explanation see text.

overshooting its host-plant by a further behavioural response to chemical odours which results in the insect turning back if the stimulating odour begins to fade.

Through evolutionary time different plant species have evolved their own specific array of secondary compounds which give them their characteristic odour and it is to this that the associated insect may respond. Polyphagous insects are attracted to plant species having similar odours. For example, larvae of the cabbage root fly *Delia brassicae* feed on a variety of Cruciferous plants and the important attractants for ovipositing females are the 'mustard oils' particularly allylisothiocyanate. Attraction occurs over a distance of up to 24 m [66]. Similar responses involving a wide range of chemicals, both separately and in combination, have been recorded for a number of different insects [67].

Orientation to the host-plant may be still more complex and involve a number of complementary stimuli. Light intensity, colour, humidity and volatile chemicals all appear to play a role in attracting the leafhopper *Empoasca devestans* to its host plant, cotton. The non-specific stimuli such as background light intensity and colour act over a longer distance than the more host specific chemical stimuli [68].

Attraction brings the insect into contact with a potential host-plant but whether that insect will settle or become arrested on that plant and commence feeding or egg laying again depends on its response to host-plant stimuli. Optical or physical stimuli such as leaf colour, leaf surface texture or leaf hairyness may determine initial acceptance but the chemical cues largely determine final acceptance. Odour again may be important as in the moth *Manduca sexta* where the stimulus for oviposition is volatile chemicals emanating from tomato leaves. However, direct contact chemoreception involving non-volatile plant chemicals is usually of overriding importance. There is wide variation in the way insects perceive and react to these chemicals and it is difficult to generalize. Some insects are 'arrested' by chemical stimuli received through contact chemoreceptors situated on the fore-tarsi or mouthparts, others immediately take a test bite. Biting brings the chemoreceptors on the mouthparts into direct contact with the plant sap. Continued feeding is dependent on the presence, at acceptable concentrations, of chemicals (phagostimulants) which stimulate feeding and the absence of chemicals which deter feeding (deterrents).

Plant nutritive substances and secondary compounds can act as both phagostimulants and deterrents. In the mustard beetle *Phaedon cochleriae* the secondary glucoside sinigrin [69] stimulates continuous feeding whereas in the grass-feeding *Locusta migratoria* nutritive hexose and disaccharide sugars are the main phagostimulants [70]. Deterrents are usually secondary plant compounds and their role in plant defence has already been discussed. A wide range of such chemicals are known to deter feeding in *Locusta migratoria* but the alkaloids and monoterpenoids are the most effective, particularly at low concentrations [71]. Nevertheless, high concentrations of certain salts, sugars and amino acids may also deter feeding in some insects [72].

Ovipositional stimuli may be equally varied and we have already noted the importance of olfaction in *M. sexta*. Oviposition by the cabbage white butterfly *Pieris brassicae* occurs when the fore-tarsi are in contact with sinigrin, whereas in some crickets oviposition is preceded by a test bite [73].

3.2 Finding the food: synchronization with the host-plant

In general, feeding, growth and reproduction of herbivorous insects can only occur when the host-plant is actively growing. Natural selection should, therefore, act to ensure a high degree of temporal synchrony between insect and plant life cycles. This is best illustrated by an example from the low arctic where the growing season is short (2.5 months) and where the problems of synchrony are greatly accentuated. In northern Alaska, dwarf deciduous willows are a characteristic component of the vegetation. Two common psyllids *Psylla palmeni* and *P. phlebophyllae* complete their development feeding on female willow catkins [74]. Fig. 3.2 compares the development period of the psyllid and the period during which catkins remain suitable for nymphal development. Almost perfect temporal synchrony is required if the psyllids are successfully to complete their life cycles.

The problem of synchrony is not confined to regions with alternating winter–summer periods. In some tropical rain forest areas many insects such as the Homoptera feed preferentially on new flushes of plant growth and others, such as the bruchid beetles, attack the fruits and seeds of specific trees. Seasonal patterns of rainfall may result in seasonal cycles of leaf flushing and fruit production. On Barro Colorado Island, Panama, an area with alternating 'wet' and 'dry' seasons tree flushing occurs predominantly during the wet season. This generally coincides with a peak in the abundance of Homoptera, again suggesting a close synchrony between insect life histories and host-plant phenology [75].

3.3 Insect feeding mechanisms

Herbivorous insects have evolved a variety of different feeding mechanisms for exploiting plant tissue and the structure of their mouthparts is well documented in the literature [76]. They can be classified into two

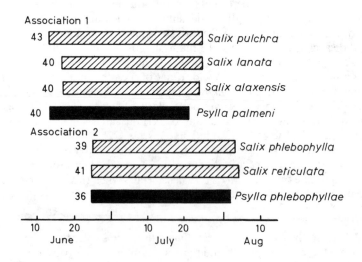

Fig. 3.2 Duration of the nymphal stages of *Psylla palmeni* and *Psylla phlebophyllae* in days compared with the periods of availability of the catkins of their respective host-plants (*Salix* spp.). (Reproduced in modified form from Hodkinson *et al.* [74], by permission of Blackwell Scientific Publications.)

main types, those adapted for biting and chewing whole plant tissue and those adapted for piercing the plant and sucking the sap. The former type is found within the important orders Orthoptera, Hymenoptera, Coleoptera and larval Lepidoptera, the latter within the Hemiptera and in a less complex form in the Thysanoptera. Other modifications occur but are less common. For example, the larval mouthparts of many phytophagous Diptera, such as the leaf-mining Agromyzidae, are reduced to a single small curved cutting or rasping hook attached to the cephalopharyngeal skeleton.

The mouthparts of chewing insects consist essentially of three pairs of appendicular jaws, the anterior mandibles, the maxillae and the posterior lower lip or labium in which the paired appendages are fused along the midline [77]. The maxillae and labium bear paired segmented sensory appendages, the palps. Associated with the mouthpart structures are an anterior lip or labrum and a median tongue-like structure the hypopharynx. The mandibles are primarily involved in the cutting and maceration of food while the maxillae are used for food manipulation in addition to aiding the maceration process. The chemoreceptor sensilla involved in host-plant recognition are variously situated on the inner labrum, on the hypopharynx and on both the main lobes and palpi of the maxillae and labium. Most chewing insects are non-selective in the way they feed; they normally ingest macerated whole leaf tissue. Some, however, are more selective; the larvae of leaf mining Lepidoptera may consume the inner tissues of a leaf, leaving the less palatable outer cuticle intact.

In sap-sucking insects such as the Hemiptera, the basic mouth parts have undergone considerable modification [78]. The mandibles and maxillae are drawn out into long, thin, needle-like stylets which fit together to form a stylet bundle, consisting of an outer pair of mandibulary stylets and an inner pair of interlocking maxillary stylets. The stylet bundle lies within an anterior groove of the labium which, itself, is extended in the form of a proboscis. The maxillary stylets enclose a food canal along which liquid sap is taken up from the plant and a salivary canal through which saliva is pumped into the plant. Innervation of either or both the mandibulary and maxillary stylets suggests that the insect is able to receive chemosensory and/or tactile information from the very apex of the stylets [79].

The development of sap-sucking mouthparts has permitted a considerable degree of sophistication in the choice of plant tissue selected for feeding. Some leafhoppers (Cicadellidae) which feed on the contents of mesophyll cells have short stylets with barbed apices [80]. These are inserted in rapid thrusts and with some lateral movement which results in the rupture of cells and the release of the soluble contents. In contrast, other Hemiptera feed on the deeper vascular tissues, the phloem and xylem. Aphids and scale insects (Coccoidea) often select phloem tissue and its location requires long, thin stylets and a considerable degree of control over the direction and depth of stylet penetration. On an

29

herbaceous plant such as *Vicia faba* the bean aphid *Aphis fabae* may take up to one hour to insert its stylets into the phloem, but the woolly aphid *Eriosoma lanigerum*, feeding on woody apple twigs, may take longer than a day fully to insert its stylets [78]. However, once the translocation stream of the plant has been tapped a continual supply of food is guaranteed without the need to change feeding site. Spittle bugs such as *Philaenus spumarius* which feed on xylem sap have a similar advantage [81].

Chewing insects remove plant tissue directly and, with the exception of some highly specialized gall-forming groups, there is little opportunity for subtle disruption of the host plant metabolism. This is not so in the sap-sucking insects where salivary injection may cause major internal disruption of the plant. Damage to plants resulting from feeding by the Hemiptera ranges from simple necrotic spots surrounding feeding punctures to gross tissue malformation and gall formation. Symptoms appear to be determined partly by type, age and physiological state of the tissue attacked and partly by the composition of the saliva injected into the plant.

Hemipteran saliva contains several enzymes, their nature reflecting the feeding habits of the species concerned. Phloem-feeding bugs usually possess carbohydrases, particularly amylase and a pectin hydrolysing enzyme involved in the breakdown of the middle lamellae of the cell walls. In addition, proteinases, esterases and lipases occur commonly in species feeding on mesophyll tissue or seeds [82]. A number of other compounds have also been isolated including metabolites such as amino acids and phenolic compounds, the plant growth regulating hormone indoleacetic acid (IAA) and the oxidizing enzyme polyphenol oxidase. The role of injected IAA as a causal agent of growth distortion in plants still remains unclear. It is uncertain whether the amounts present in the saliva are sufficient to produce the observed effects or whether they are produced by other components of the saliva. These may interfere with the IAA-oxidase system by which the plant controls its hormone balance. The role of the polyphenol oxidase enzyme is similarly poorly understood, although in the rose aphid *Macrosiphum rosae* it is thought to be involved in the detoxification of the phenolic compounds catechin/epicatechin [83].

Another feeding characteristic of the Homoptera and the Heteroptera: Pentatomorpha is the secretion of a salivary sheath, to form an inert and impermeable proteinaceous coat around the stylets. This remains embedded within the plant when the stylets are withdrawn, serves to attach the mouthparts to the plant during stylet penetration and acts as a sleeve to the stylet 'bore hole.'

3.4 Food utilization and conversion efficiencies

During their development, insects pass through a number of larval stages or instars. Growth between successive instars is logarithmic and the amount of food consumed follows the pattern shown in Fig. 3.3.

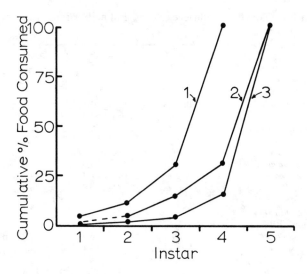

Fig. 3.3 Utilization of total food required by an insect to complete development shown as the cumulative percentage consumed by successive instars. 1. *Phytodecta pallidus* (Coleoptera) [98]; 2. *Chorthippus dorsatus* (Orthoptera) [87]; 3. *Pieris brassicae* (Lepidoptera) [84].

Usually over 60% of the total food required for larval development is consumed during the final instar.

A series of energy balance equations have been derived which serve as a useful means of investigating the efficiency of food utilization by insects [85]. They are:

$$MR = NU + C$$
$$C = P + R + FU = A + FU$$

where MR is the food energy removed by insect, C is the food energy consumed, NU is the food energy not utilized, P is the food energy going to production of body tissue, A is the food energy assimilated, R is the food energy respired, and FU is the food energy passed out as faeces and urine. From these equations we can derive the important conversion efficiencies A/C, P/A and P/C, where A/C is the proportion of food energy consumed which is assimilated, P/A is the proportion of assimilated energy going to tissue production and P/C measures the conversion efficiency of energy consumed into body tissue.

Some chewing insects are inefficient feeders, removing from the plant more tissue than they actually consume. The grasshopper *Chorthippus parallelus* consumes only 59% of material removed, the rest being dropped. Tropical leaf cutting ants belonging to the genera *Atta* and *Acromyrmex* can remove large quantities of leaf material from trees. This is not consumed directly but is transported back to the nest and used as a substrate for the cultivation of fungi.

Table 3.1 Food conversion efficiencies of typical herbivorous insects expressed as a percentage

Species	Tissue	A/C	P/A	P/C	Reference
Chewing insects					
ORTHOPTERA					
Chorthippus parallelus	Leaves	34.5	42.5	15.0	[86]
C. dorsatus	Leaves	68.0	34.0	23.0	[87]
Encoptolophus sordidus	Leaves	26.1	48.6	12.7	[88]
Melanoplus sanguinipes	Leaves	57.1	50.7	29.0	[89]
LEPIDOPTERA					
Hyphantria cunea	Leaves	30.0	55.0	17.0	[90]
Pachysphinx modesta	Leaves	41.4	46.0	19.0	[91]
Hydriomena furcata	Leaves	41.8	38.8	20.0	[92]
Oporinia autumnata	Leaves	37.0	45.9	17.0	[93]
Phragmataecia cataneae	Stem	25.1	73.3	18.4	[94]
HYMENOPTERA					
Dineura viridotarsa	Leaves	16.0	56.3	9.0	[93]
Neodiprion sertifer	Leaves	13.4	60.4	8.5	[95]
DIPTERA					
Hedriodiscus truquii	Algae	59.0	30.0	17.8	[96]
COLEOPTERA					
Leptinotarsa decemlineata	Leaves	45.4	61.0	27.7	[97]
Phytodecta pallidus	Leaves	50.0	57.8	28.9	[98]
Anomola cuprea	Roots	19.6	49.0	9.6	[99]
Callosobruchus analis	Seeds	85	58	50	[100]
Sap-sucking insects					
HEMIPTERA					
Leptopterna dolabrata	Mesophyll cells	32.5	56.4	17.8	[28]
Cicadella viridis	Xylem sap	47.3	28.1	13.3	[101]
Neophilaenus lineatus	Xylem sap	41.6	36.9	15.4	[102]
Strophingia ericae	Phloem sap	22	51	11	[103]
Macrosiphum liriodendri	Phloem sap	33.4	80.6	27.0	[104]

Table 3.1 shows some typical efficiencies for a range of insects feeding on different plant tissue. There is a great deal of variability even within particular groups which is attributable in part to different experimental conditions and methods of measurement. However, the energy content of the food is not necessarily a good measure of its nutritive value for a growing insect and we might expect the above values to vary, according to food quality, particularly the available nitrogen levels. Such variation has been demonstrated both between and within host plant species. For example, larvae of the moth *Operophtera brumata* have an A/C ratio of 18 and 26% on hazel and oak respectively [105]. Similarly there are marked differences in the pattern of energy utilization by the aphid *Aphis fabae* feeding on the young and mature growth of bean plants [106]. There appears to be a broad correlation between the nutritive value of the food measured as the available nitrogen content and the

efficiency with which the food is assimilated [24]. The higher values in Table 3.1 are for insects feeding on algae, seeds and young succulent leaves: the lower values are for insects feeding on woody tissues, old mature leaves and low quality plant sap. Changes may also occur in the efficiency values as an insect passes through its life cycle [107].

Some insects may be able to compensate for low quality food by increasing their consumption rate. When larvae of the butterfly *Pieris rapae* are fed on plants with different nitrogen levels they appear to adjust their food intake and assimilation rate to stabilize the rate of nitrogen accumulation. On plants with a low available nitrogen level they eat more and assimilate nitrogen more efficiently than on plants with higher nitrogen levels [108]. Foliar water content may also play an important role in insect nutrition. Experiments on a number of Lepidoptera species have shown that the assimilation efficiency of leaves of similar nitrogen levels is influenced by leaf water content. Leaves are utilized less efficiently as water content falls. This effect is more pronounced in species feeding on trees as opposed to herbs [109].

In some Hemiptera, notably the aphids and psyllids, the density of feeding insects themselves may be an important factor determining the amount and quality of available food. As these insects feed, saliva is injected into the plant and the cell tissues are broken down. Increasing insect densities, up to an optimum, may lead to an improvement in the nutrient supply to the individual which results in an improved rate of growth or survival [110, 111, 112]. Fig. 3.4 shows the survival of the Australian eucalyptus psyllid *Cardiaspina densitexta* on leaf discs. This reaches an optimum at a density of ten nymphs per disc.

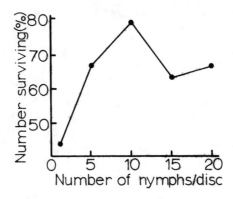

Fig. 3.4 Survival of the nymphs of *Cardiaspina densitexta* at different densities on leaf discs. (Reproduced from White [111] by permission of CSIRO.)

33

4 Insect herbivory and non-woody plants

Herbaceous plants are usually relatively short-lived and lack the resistant structural materials found in the woody plants. Thus, the whole plant is potentially susceptible to insect herbivory. Much of the literature dealing with the effects of insects on such plants relates to agricultural crops and there is little detailed information on non-economic species. Nevertheless, ideas derived from insect–crop plant relationships are generally applicable and we can examine the relationship at both the individual plant and the plant population level. Many of the ideas presented are also applicable to the woody plants such as trees (see Chapter 5) but because of their larger size and longer lifespan trees are more difficult to study.

4.1 Herbivory and the individual plant

At first sight the effect of a chewing insect on a plant might appear simple: there is an immediate, measurable loss of leaf area and an equivalent drop in the plant's photosynthetic capacity. The relationship between leaf damage and plant productivity is, however, complex and depends on several interrelated factors.

First, there is the plant itself, with its complex growth pattern and capacity to translocate material between tissues. We can view a growing plant as a number of interlinked *sources* and *sinks*. The sources are the plant organs, primarily the leaves, involved in the synthesis of food materials. The sinks are the organs such as the growing shoots, roots, storage organs and reproductive structures to which food produced by the sources is translocated.

During the life of a plant the contribution of a single source, such as an individual leaf, to the overall economy of the plant will vary greatly. Similarly, certain sinks such as the reproductive structures will only operate at a given stage of the plant's development. For example, the early leaves of wheat plants do not contribute directly to the growth of the ear: it is the flag leaf lying immediately next to the ear which contributes to grain development. Thus, the aphids *Sitobion avenae* and *Metopolophium dirhodum* only reduce grain yield when feeding on the flag leaf [113]. It is clear, therefore, that the actual site of insect feeding, whether it be a source or a sink, will govern the effects on the plant and the plant's capacity to respond. Furthermore, the type and age of the plant tissue itself may determine its palatability and hence susceptibility. For example, the European corn borer *Ostrinia nubilalis* tunnels in the stem of field corn (*Zea mays*). Infestation near the base of the plant has a greater effect on ear yield than infestation at higher internodes. The larvae, however, grow better on the younger apical internodes [114].

Timing of insect attack in relation to the stage of development of a plant may also be important, especially in plants with little storage material in the propagative tissue. Wheat seedlings are more easily destroyed by the stem-boring wheat bulb fly *Leptohylemyia coarctata* at the one shoot stage than at the later two or three shoot stage [115].

The effect of feeding by some colonial Homoptera, particularly aphids, is to create their own metabolic sinks which compete with the normal sinks for the supply of nutrients. Experiments using radioactively labelled CO_2 fed to plants have shown that colonies of the aphid *Brevicoryne brassicae*, feeding on the leaves of brussels sprouts plants, increased the rate of nutrient flow into the infested leaves and decreased the flow into growing tissues [110].

Individual plants are often able to compensate to some degree for losses to insects. Compensation occurs most commonly when an insect feeds on a source which is manufacturing more photosynthate than the sinks can use. Experimental defoliation experiments on young sugar beet in which up to 50% of the foliage was removed showed no detectable change in the yield of the developing beet [116]. Direct competition for photosynthate between sinks may also occur and insects feeding on one sink may merely release another sink from competition. For example, individual wheat plants at normal field densities produce more shoots than survive to produce ears. Insects such as *L. coarctata* may kill some shoots, thereby releasing others from competition.

In addition, quantitative changes in insect feeding may induce parallel qualitative changes in plant tissue. The meadow capsid bug *Leptopterna dolabrata* feeding on wheat reduces the yield of grain but also alters its quality. The starch and gluten content is unaffected but the protein level rises with increasing insect infestation [117].

Background environmental conditions such as soil fertility levels and soil moisture potential may, through their influence on plant vigour, determine a plant's susceptibility to insect attack. Wheat plants grown on soils with adequate potassium levels are better able to withstand a given level of attack by *L. coarctata* than plants grown on potassium deficient soils [118]. There is some evidence that fertilizing plants, particularly with nitrogen may actually increase their acceptability to insects, with a resulting increase in insect reproduction. The aphid *Brevicoryne brassicae* shows such a response when fed on brussels sprouts plants fertilized with nitrogen [119]. This must, however, be set against the increased vigour of the fertilized plants which makes them more resistant to insect attack.

Insects feeding on root tissue can directly alter the capacity of a plant to take up mineral nutrients and water from the soil. Larvae of the scarabaeid beetle *Sericesthis nigrolineata* feeding on the roots of rye grass reduce the growth of both new roots and new foliage and can

36

induce symptoms of water stress in the plants [120].

The overall effect of insect grazing will ultimately be determined by the insect population density and the feeding pressure it is able to exert. This will not remain constant but will change throughout the life of the plant.

4.2 Herbivory and the plant population

So far we have examined the effect of insect populations on individual plants. However, plants rarely grow in isolation and ultimately we are concerned with the effects of herbivory on plant populations.

The spatial distribution and density of the plant population itself may play an important role in determining the level of insect attack. The relationship is highly complex, being determined by an interplay of factors. These appear to include the physical structure of the vegetation and its associated microclimate, the influence of parasites and predators and the host-finding and reproductive behaviour of the insect [121]. Often the highest densities of insects per plant are associated with the lower plant densities [122]. However, in studies on the effect of stand density on the herbivores associated with soya-beans, the numbers of leafhopper *Empoasca fabae* per plant were highest at lower plant densities. In contrast, the thysanopteran *Sericothrips variabilis* was more abundant at high plant densities [123]. Thus plants of the same species growing at different densities may differ in their susceptibility to insect herbivory. However, in a study on the effect of feeding by the Lepidopteran *Battus philenor* on its host plant *Aristolochia reticulata* growing at different densities, no differences in plant seed production could be demonstrated, despite high levels of defoliation [124].

Another major potential outcome of insect herbivory is to alter the competitive fitness of plant populations at both the intra- and inter-specific level. Take again the example of wheat growing at normal planting densities: individual plants yield about 2 g of grain. Grown in isolation the same plants will yield up to 50 g: the difference is due to competition between plants and when this is removed a large increase in yield can be expected. Infestations of wheat bulb fly kill young plants but the population may compensate for this by increasing the yield of individual plants [125].

Plant populations often contain genotypes with different susceptibilities to insect herbivory. For example, natural populations of wild cabbage contain some plants which attract more ovipositing butterflies (*Pieris brassicae*) than others. Thus, fewer larvae develop on the less attractive plants [126]. Traditionally, such differences have been exploited by breeding insect resistant plant varieties. When plant populations containing resistant and susceptible genotypes are subjected to herbivory there may be an overall shift in the competitive abilities of the two genotypes. Varieties of barley differ in their susceptibility to the grain aphid *Schizaphis graminum*. When a resistant and a susceptible barley variety were grown in competition, the susceptible variety was the

37

better competitor. However, when the cultures were exposed to aphid feeding the outcome was reversed: the resistant variety became the better competitor [127].

Plant populations are also subject to interspecific competition and it is reasonable to assume that herbivory may cause changes in the relative competitive abilities of competing plant species. Ultimately, this will result in changes in community composition. This will be explored fully in Chapter 6.

4.3 Quantitative relationships

The relationship between plant yield or net productivity and insect feeding, measured as number of insects or feeding injuries caused, can be described by the Tammes' response curve (Fig. 4.1) [128]. This is a generalized curve which describes in quantitative terms the phenomena already discussed. It can be used to describe the response of a plant organ, an individual plant or a population of plants to insect feeding. Nearly all plant–insect associations display at least some of the features illustrated by the curve which can be divided into three distinct parts.

At low insect population density the plants may be able to compensate completely for damage and there is no reduction in net production. At slightly higher insect densities, beyond a threshold level, compensation becomes less effective and production begins to decline. The curve then straightens, indicating a linear relationship between yield loss and increasing insect density. Beyond this point competition for food between the insects may reduce their individual effectiveness, resulting in a gradual decrease in the gradient. In some cases a lower plateau may

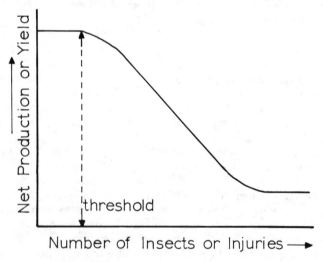

Fig. 4.1 Generalized curve to show relationship between the yield of a plant and the number of insects or injuries. (Reproduced from Tammes [128], by permission of Netherlands Society of Plant Pathology.)

occur such as when insects feed on the leaves of plants that have underground storage organs. Material previously accumulated in these organs will not be destroyed even by complete above-ground defoliation.

For some insect–plant associations, particularly those involving aphids, a better fit at least to the middle section of the curve is obtained if the insect population density is expressed in logarithms [129]. Experimental data are often variable and approximate fits are the best that can be obtained. Fig. 4.2 shows the results of a typical experiment designed to measure the impact of the aphid *Macrosiphum avenae* on the grain yield of wheat. In this instance there is little evidence of either compensation or of a lower threshold [130].

Insect feeding usually results in a loss of plant production but there are examples, admittedly uncommon, where feeding can stimulate plant growth [131]. Larvae of the moth *Plutella xylostella* feed preferentially on the young leaves of turnip and stimulate the plant to retain older leaves which are normally shed. This results in an increase in the total dry matter produced [129]. Similarly, small colonies of *Aphis fabae* on field beans appear to suppress apical growth thereby stimulating an increased production of beans [132].

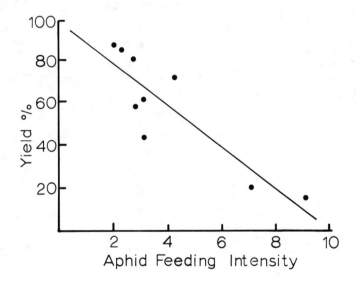

Fig. 4.2 Relationship between the yield of wheat and aphid feeding intensity measured on a relative scale. (Based on data from Rautapää [130].)

5 Insect herbivory and woody plants

Wood is persistent and woody plants may live to a great age. An important reason for this is that relatively few insects have evolved to exploit wood in living plants. Most species feed on the softer and more transient tissues of woody plants such as leaves, fine roots, flowers, fruits and to a lesser extent the meristems. The primary effect of insect herbivory is on the photosynthetic, nutrient winning and reproductive mechanisms of the woody plant, rather than on the persistent woody support, transport and storage organs. Leaves, flowers and fine roots are often replaced after they have been consumed. This may take place during the same growing season or over a period of two or more years, depending on the growth pattern of the plant species concerned. The repair and replacement of damaged tissue involves a diversion of resources which often results in a reduction in the growth rate of other tissues. Plants with substantial perennial storage organs can, however, withstand considerable losses of production to insects over short periods.

Meristematic tissues are a further potentially useful food source for insects but damage to them is potentially less reversible than damage to leaves or flowers. For example, relatively minor injury to the apical meristems of young trees may inhibit their height growth and reduce their ability to compete for light in the forest understorey. Fewer insects appear to have adopted the higher risk strategy of feeding on meristems.

In this chapter we examine the spatial patterns and levels of consumption by insects in a forest canopy and investigate the effects of insect feeding on the growth of woody plant species.

5.1 The distribution and intensity of insect herbivory

A tree canopy is a complex structure and foliage is not distributed uniformly within it. The leaves of hardwoods are often smaller at the top of the canopy than at the bottom whereas the converse applies in some conifers. The date of bud opening and the rate of leaf development may differ considerably within and between trees of the same species. Furthermore, there may be structural and physiological differences between sun and shade leaves. Thus the extent and impact of insect herbivory may differ spatially and temporally both within and between trees of a given species. For example the feeding activity of leaf eating insects in Danish beech forests is concentrated in the lower canopy [133].

A detailed comparison of the rates of plant consumption by insects in different forest ecosystems is reserved until Chapter 7 but it is appropriate at this stage to discuss the more general problems. In the north temperate zone, it appears that leaf-feeding insects are, for most of the type, typically engaged in low levels of herbivory. For example, an eight year study on Danish beech (*Fagus sylvatica*) forest showed that loss of leaves to the dominant herbivores, the weevils, *Rhynchaenus fagi* and *Phyllobius argentatus*, ranged from 7 to 20% of leaf area, averaging 12.3%. After correcting this figure to allow for the growth of leaf hole boundaries and leaf necrosis, the percentage consumption on a dry weight basis represents no more than 5% of the available food, the leaf lamina [134]. In contrast, sap-sucking insects utilize a different, if related resource and it is difficult to obtain reliable estimates of their consumption. Not unnaturally, intensive studies have been made on tree species which commonly support large populations of sucking insects. For instance, it has been estimated that lime aphids (*Eucallipterus tiliae*) divert photosynthate equivalent to 19% of the net primary production of mature lime trees [135]. It is unwise, however, to extrapolate this figure to other tree species and we must conclude that the typical levels of consumption by these insects remain uncertain.

From time to time, leaf eating insect populations increase explosively to produce an outbreak when most or all of the leaf canopy may be eaten. Such outbreaks are relatively uncommon but their effects may be highly significant, particularly in commercial forestry. The literature on the effects of insect defoliators on trees is strongly biased towards the documentation and assessment of these outbreaks [136]. The frequency of outbreaks varies greatly. Studies on the Lepidoptera of Scandinavian birch forest have revealed twelve outbreak periods of *Oporinia* and eight of *Operophtera* with outbreaks lasting an average of 7 years and with a 9.5 year interval between the first years of successive outbreaks [1]. Similar synchronous outbreaks of tussock moth (*Orgyia pseudotsugata*) larvae on conifers occur every 8–10 years over extensive areas of western North America [137]. In contrast such regular and extensive eruptions do not occur in British Oakwoods.

5.2 The consequences of herbivory for the woody plant
Woody plants are complex integrated organisms and canopy damage does not necessarily produce proportional changes in stem height and diameter growth, seed production or root extension. The effect of insect feeding on a particular tree organ may be transmitted to other organs and may ultimately affect the growth of the whole tree either in the current or in subsequent years [138]. Furthermore, tree species may differ significantly in the way they respond to the energy and nutrient drain imposed by insects and this often makes it difficult to generalize. Studies on the effects of the sycamore aphid *Drepanosiphum platanoides* and the lime aphid *Eucallipterus tiliae* on their respective hosts have shown that both reduce growth but this is brought about in different ways. In sycamore there is a reduction in shoot and root growth, a decrease in leaf size, but an increase in leaf chlorophyll content. The leaves are able partially to compensate for losses of photosynthate to the

aphids by increasing the rate of carbohydrate production. In contrast, lime trees show no reduction in leaf size or shoot growth but root growth is seriously impaired. Aphid infested leaves contain lower chlorophyll levels and are unable to compensate by increasing their rate of photosynthesis. However there is a lag effect between years. When aphids are absent in the subsequent year, the newly produced leaves contain higher chlorophyll levels and fix more energy than leaves from previously uninfested trees [139].

Trees are large organisms which live for a long time and it is difficult to conduct simple experiments which measure the effect of insect feeding on their growth. There are several thousand reports which associate insect herbivory and damage to trees, but very few give reliable estimates of both insect activity and the consequences for the tree. Three main approaches have been used to determine the response of trees to insect feeding and each provides slightly different information. The first approach is to observe the response of trees to known levels of artificial defoliation. The second is to exclude insects from selected trees and compare the growth of these trees with ones subject to normal insect feeding. The third approach is to make repeated long term measurements of insect population density or levels of defoliation and correlate these observations with parallel measurements of tree growth.

5.2.1 Experimental defoliation

Artificial defoliation experiments have, for practical reasons, usually been conducted on young trees. Considerable care has to be taken in the experimental design as growth rates can vary between individuals of identical genetic origin and minor variations in microsite conditions may produce significant effects on growth. Artificial defoliation may be applied in various ways to simulate the effects of insect feeding. The proportion of leaves removed, the date and number of defoliations and the location of the treatment within the canopy can all be varied. Individual leaves can be removed completely or in part. Leaves of different ages in evergreen plants may be treated differently.

An experiment was conducted on the American elm (*Ulmus americana*) in which small trees were completely defoliated once, in early June, or twice, in early June and late July [140] for two successive years. Die-back of shoots was greatest in the biannually defoliated trees but even after two years only one out of fourteen trees had died. The weight of twigs on defoliated trees was about one-third that on controls while stem diameter growth was reduced to about one-quarter. Recovery from defoliation in the annually defoliated trees took place by further terminal growth on each twig whereas the second annual defoliation induced growth from axillary buds. New leaf size was reduced with successive defoliations although leaf number responded in a more complex way. Growth in the following year was affected both by delays in bud-opening and by the development of smaller buds on shoots formed after defoliation.

Similar experiments have been conducted on red oak (*Quercus rubra*) and red maple (*Acer rubrum*) in which trees had 50%, 75% or 100% of their leaves removed in late June over three successive years [141]. Regeneration of the leaf canopy took place within three weeks at 100% defoliation and four to five weeks at 75% defoliation. However, only some trees reflushed after 50% defoliation. Successive defoliation of red oak produced earlier budbreak the following year. In contrast, budbreak in red maple was delayed by 100% defoliation but advanced by 50% and 75% defoliation. Leaf growth in succeeding years was most affected by the first defoliation. The carbohydrate composition of the primary leaves, those formed prior to the annual defoliation, was affected by the severity of previous defoliations. Leaf carbohydrate declined with increased level of defoliation except in red oak where the 50% defoliated trees suffered the greatest reduction. Nitrogen content of primary leaves, however, was unaffected in both species. The most striking outcome of these experiments was the lack of evidence for any straight-forward cumulative effect of defoliation in successive years. It was the first defoliation which had the greatest impact. This suggests that like the American elm, red oak and red maple have remarkably effective recovery mechanisms.

It is interesting to compare these results for deciduous trees with data for similar experiments on Scots pine (*Pinus sylvestris*) which normally retains its needles for several years [142]. The more severe and persistent the defoliation of Scots pine, the greater was the reduction in the diameter growth, the length of shoots, the needle biomass and the number of buds. Late summer defoliation generally produced greater effects the following year than early summer defoliation. There was, as others have found for conifers, a delay of up to two years after the removal of current needles before the most severe effects on stem diameter growth were revealed [143, 144]. The main organs of the tree were affected to different extents. The order of decreasing susceptibility was: stem diameter growth > shoot length > needle biomass > bud numbers. Starch reserves in the needles of defoliated trees were depressed, particularly when current year needles were removed in late summer. Coniferous trees build up starch reserves in their needles before bud-break and this material is subsequently mobilized to support growth [145]. The reduction in the availability of carbohydrate probably contributes to the general reduction in growth, although other factors such as mineral nutrients and plant hormones are likely to be important. The results suggest that the tree coped with defoliation by giving needle production priority over stem and shoot growth. The best strategy for recovery from partial or complete defoliation seems to be to produce as nearly normal a photosynthetic machinery as possible. This appears to be a common response, particularly in those deciduous trees capable of multiple leaf flushes, such as the red oak and red maple. The tree is thus able to take advantage of its perennial nature and survive almost intact to more favourable times.

5.2.2 Insect exclusion experiments

Artificial defoliation experiments can indicate the way in which a tree might respond to high levels of defoliation by leaf-chewing insects. An alternative and perhaps more realistic experimental approach is to use insecticide treatments to exclude insects from the plants under study. The growth of sprayed plants can then be compared with unsprayed controls. This method measures the combined impact of the total insect fauna.

An exclusion experiment of this type has been conducted on newly established plots of broom (*Cytisus scoparius*) over an eleven year period [146]. One experimental plot was repeatedly sprayed with insecticide while a control plot was left unsprayed. The unsprayed plot was colonized more rapidly by insects, both in terms of number of species and individuals than was the sprayed plot. Almost half of the 240 bushes planted on the unsprayed plot died within ten years, while on the sprayed plot less than one-quarter died. The unsprayed bushes produced only one-quarter the seeds of the sprayed bushes over the same period. At the end of the experiment, broom bushes in the unsprayed plots were about 77% the height of the sprayed bushes. Unsprayed bushes were stunted and bushy with shorter internodes, an effect often seen in woody plants attacked by psyllids and aphids. The changes in growth rate, mortality, natality and growth form observed in this experiment resulted not only from a reduction in plant's photosynthetic machinery by insects but also from other effects such as seed predation, disease transmission and disruption of the plants' nutrient and hormonal balance. Broom may, however, be somewhat atypical as even in older bushes the proportion of inedible 'wood' to edible 'green' tissue is about 1.5:1. Thus, relatively more of the broom is available to insect herbivores than is often the case for forest trees where the ratio may be as high as 20:1 [147].

Eucalypt species in Australia experience high levels of insect herbivory. Exclusion experiments have been conducted on two species *Eucalyptus pauciflora* and *E. stellulata*, both of which are multistemmed [148]. The experiment involved treating one stem per tree with insecticide for one year and comparing the growth of these stems with matched unsprayed stems from the same trees and with stems taken from other unsprayed control trees. The sprayed stems of both species showed diameter growth up to 2 to 4 times greater than the controls. The unsprayed stems on the 'sprayed' trees showed a more modest relative increase. The effect of spraying on tree growth lasted at least two years, but thereafter declined. If these *Eucalyptus* species have the same response strategy as the oak and pine discussed in the previous section, the impact of insect exclusion on the growth of the other plant organs may well be less than on diameter growth. However, it is well known that eucalyptus planted in other parts of the world on impoverished soils without their endemic insect fauna, show remarkably high growth rates

[149]. This suggests that their growth is significantly suppressed by insect herbivore activity.

5.2.3 Collation and correlation of observations

The impact of insect herbivory on tree growth can be investigated by collating observations on intensity of defoliation and tree growth. This approach is perhaps the most natural as it does not involve experimental manipulation of either the plant or the insect population. It suffers the disadvantage that observations must be made over long periods before realistic conclusions can be drawn. There have, in consequence, been few studies in which the insect population, the level of defoliation and the growth of the tree have been observed simultaneously for a long time.

In a study of winter moth (*Operophtera brumata*) caterpillars on red oak (*Quercus rubra*) in Nova Scotia, percentage defoliation, tree mortality and stem radial increment was measured over a period of years [150]. The mortality and growth data were compared with those from undefoliated trees. Increasing defoliation was accompanied by increasing tree mortality and a loss of leaf production. Radial increment in undefoliated trees was 23% over four years but this decreased proportionately to 8% at a cumulative defoliation of 250%.

The problems associated with this type of investigation are well illustrated by the results of a twenty year study of caterpillar populations and the growth of five oak trees at Wytham Wood, England. This study was first reported on the basis of eight years' data [152]. Variation in mean caterpillar density was shown to account for 79% of the variance in latewood growth, expressed as a percentage mean. Extrapolation back to zero caterpillar numbers suggested a 60% loss of latewood growth resulting from caterpillar activity. These figures were later recalculated for twenty years' data [151]. Caterpillar density then explained only 48% of the variation in latewood growth and the slope of the relationships was less than one-third as steep (-0.22 as compared with -0.74). The estimated loss of latewood growth fell to near 20% and was statistically non-significant. Thus, in this case, firm conclusions cannot be drawn safely, even from twenty years' data.

5.3 Other effects of insect herbivory

Insect herbivory affects not only the rate of growth and the functioning of the photosynthetic machinery. The effects may be more subtle. The impact of the balsam woolly aphid (*Adelges piceae*) on grand fir (*Abies grandis*) has been particularly well documented [153, 154]. Aphids cause a disturbance of tree metabolism which results in a large reduction in the carbohydrate reserves of needles and twigs. Obvious effects are twig deformation and increased tree mortality. The more subtle effects involve the disturbance of xylem formation. The number of pit pores in each conducting tracheid is reduced to about one third that found in non-infested trees [155]. This can be interpreted as a premature conversion of sapwood to heartwood by aphid attack and appears to be related to a reduction in the water permeability of the wood [156]. It is suggested that the reduction in the water conducting capacity of the stem may result in water stress in shoots and a reduced carbohydrate build-up. Mortality of rootlets in a related species, balsam fir (*Abies balsamea*) has been linked to above-ground defoliation by the spruce budworm (*Choristoneura fumiferana*) [157]. This may have similar severe consequences for old trees with little capacity for replacing lost rootlets.

There is a great deal of evidence that trees weakened by heavy and repeated defoliation may be particularly vulnerable to attack by disease and secondary pest insects. It is often difficult to distinguish between insects that attack weakened trees and those which invade trees which have recently died [158]. Infection of aspen (*Populus tremuloides*) by the fungi *Hypoxylon* and *Nectoia* has been shown to rise with increasing severity of defoliation by the forest tent caterpillar (*Malacosoma distria*). Attack by bark-boring insects has also been shown to increase at higher levels of defoliation [159]. A physiological change in a tree is often a prerequisite to insect attack and defoliation may produce just such changes [138].

5.4 Insect herbivores and tree rings
Major outbreaks of insect herbivores can reduce or modify the production of xylem tissue in many species of tree. The record left in the annual rings can be used to reconstruct the past history of outbreaks.

Spruce budworm (*Choristoneura fumiferana*) outbreaks produce a characteristic suppression of ring width in balsam fir (*Abies balsamea*), white spruce (*Picea glauca*) and black spruce (*Picea mariana*). Ring width measurements for these host species have been compared with a matched sequence for a non-host species, eastern white pine (*Pinus strobus*) [160]. Marked divergences over two to three years have been taken to indicate an insect outbreak. A comparison of trees up to three hundred years old revealed six major outbreaks at varying time intervals.

Sometimes the structure and appearance of the annual ring is also modified by heavy defoliation. Severe damage to *Eucalyptus delegatensis* by the phasmatid *Didymuria violescens* reduces the width of the dark band of latewood formed at the end of the year of defoliation and of the earlywood formed the following year [161].

Intense defoliation of European larch (*Larix decidua*) by larch bud moth larvae (*Zeiraphera dinana*) modifies wood structure by producing thick walled or small lumened cells which can be easily distinguished from the normal bands of latewood cells by X-ray densitometry (Fig. 5.1) [162]. Samples taken from trees and timbers in Switzerland have been used to reconstruct the pattern of bud moth outbreaks over several hundred years. Fig. 5.1 compares a reconstructed history of outbreaks with observed historical records and there is almost complete agreement. Outbreaks were shown to vary both in intensity and regularity

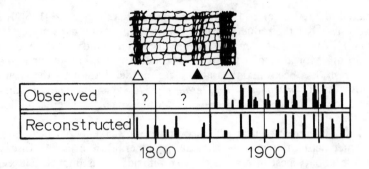

Fig. 5.1 Tree ring records of larch bud moth outbreaks. The upper figure represents a cross section of wood showing the typical bands of latewood formed in successive years (open triangles) and a band of false latewood associated with defoliation (closed triangle). False latewood has a more diffuse boundary and thinner walled cells than true latewood.

The lower figure compares the pattern of outbreaks as reconstructed from tree ring analysis with the observed historical record of outbreaks in the Engadin region of Switzerland. (Modified from Schweingruber [162], with permission.)

according to site conditions. However, samples from the Valais region revealed a remarkably uniform pattern of twelve to thirteen outbreaks per century over almost six hundred years.

6 Insect herbivory and the plant community

In this chapter we will examine how plant communities can modify, either directly or indirectly, the abundance and population dynamics of their associated insect herbivores and, conversely, how insect herbivores might alter the composition of plant communities [163].

6.1 Plant community composition and insect abundance

The distribution and physical structure of vegetation can influence the spatial patterns of insect herbivore populations. In areas of arctic Alaska five species populations of jumping plant lice (Psylloidea) are variously associated with nine different species of willow (*Salix* spp.) [74]. The area is topographically diverse with polygonized tundra, seasonal lake basin, dry ridge, bluff, sand dune and river edge habitats occurring in close proximity. Each willow species shows a characteristic overlapping distribution across the different habitat types. In some habitats such as the dry ridge, the willows are subject to intense psyllid feeding pressure whereas in others, such as the river edge, sand dunes and lake basin, certain psyllids may be completely absent. The latter *Salix* habitats are subject to seasonal perturbations such as ice movement, flooding or wind blow which prevent the psyllids establishing breeding populations.

Larger vegetation types such as trees and hedges can locally modify air movements and cause the deposition of wind dispersed insects. These insects tend to accumulate in the lee of windbreaks, although this can be modified by the species richness of the surrounding vegetation and its attractiveness to insects [164, 165, 166].

In Chapter 4 we noted that the density of insect herbivores may be influenced by the density of their host-plant. The structural diversity and species composition of the plant community in which a particular insect–plant association is found may also affect the insect's abundance. This background diversity may influence herbivore populations in three main ways [167]. First, in diverse communities the visual and chemical stimuli by which the insect locates its host may become diffuse and confused resulting in reduced success in host plant location and ultimately a lower population density. This effect occurs when the flea beetle *Phyllotreta cruciferae* feeds on collards (*Brassica oleracea*) grown in diverse culture as opposed to monoculture [168].

Secondly, increased vegetational diversity may encourage predators by providing shelter or increased numbers of alternative prey which help maintain a higher predator density. The diversity of *Brassica oleracea*

49

crops is increased by undersowing with clover. This reduces the number of eggs laid by the cabbage root fly *Delia brassicae* and the larval survival of the butterfly *Pieris brassicae*. The latter effect is associated with an increase in the numbers of predators [169].

Thirdly, the plant community in which a particular host-plant is growing may, through competitive or allelopathic effects, alter the availability of the host-plant to the insect. This may involve a reduction in plant size, a change in plant quality, or a modification of the seasonal growth pattern.

These factors will often act together or, occasionally, in opposition. Observations of the chrysomelid beetle *Gastrophysa viridula*, which feeds preferentially on broad-leaved dock (*Rumex obtusifolius*), show that habitats of different diversity support different insect populations. Lowest population survival occurs in diverse or mature habitats which support a higher predator density but in which, paradoxically, more food appears to be available. Highest survival is found in monoculture, associated with low plant densities and in mown hay fields which are maintained in an early successional state [167, 170].

Thus, despite apparent differences in the mechanisms involved there appears to be a negative correlation between the densities of insect herbivores and plant diversity. This has important implications for the design of agricultural cropping systems.

6.2 Effects of insect herbivory on plant communities

A plant species growing in a mixed community will be subject to competition from other coexisting species. Community composition will, therefore, reflect the competitive equilibrium between the species. Insect herbivores have the potential to alter the relative fitness of the competing species and bring about community change. They can feed directly on plants reducing their numbers, vigour or reproductive output or they can influence future generations by acting as seed predators or pollinators.

Little is known about these relationships and our ignorance can be highlighted by restating an example given by Harper in 1969 [171]. In the mid 1940s large areas of Californian rangeland were infested with the weed *Hypericum perforatum*. Biological control was successfully implemented by introducing the phytophagous beetle *Chrysolina quadringemina* and within a few years *Hypericum* had become an uncommon plant. Harper pointed out that if we had been unaware of past history we might conclude wrongly today that the *Hypericum/Chrysolina* association was an unimportant component of the rangeland ecosystem and that *Chrysolina* played an insignificant role in controlling the abundance of *Hypericum*. The same wrong conclusion could apply equally well to most insect–plant relationships.

A preliminary investigation has been made into the effect of feeding by the beetle *Gastrophysa viridula* on the competitive interaction between two dock species *Rumex obtusifolius* and *R. crispus*. Levels of feeding which had no significant effect on either species when grown in isolation resulted in extensive damage to *R. crispus* when the two species were competing (Fig. 6.1). *Rumex crispus* responded by reducing its root to shoot dry weight ratio from 2.14 to 1.69, making more material available for consumption whereas *R. obtusifolius* increased the ratio from 1.18 to 3.57 ensuring that material was protected from above ground feeding, within the root system [172, 173]. However, in further field experiments using mixed cultures subjected to normal beetle feeding pressure total seed production and seed weight was significantly reduced in *R. obtusifolius* whereas in *R. crispus* the number of seeds was not reduced but seed weight was lower in one experiment [174]. These results indicate a complex interaction in which herbivory can modify both the relative competitive fitness of the individual growing plants and their reproductive potential.

Insects are known to transmit a wide variety of important plant pathogens including fungi, bacteria, viruses, viroids and mycoplasmas. Much is known about such diseases in commercial crops but virtually nothing is known about their importance in natural vegetation, and their possible role as agents of change in communities [175]. The effects of insect feeding and disease may combine to alter the equilibrium between competing plant species. For example, the Australian trees *Eucalyptus dalrympleana* and *E. pauciflora* grow together in mixed species associations. In dense immature stands, where inter-plant competition is presumably intense, the normal ratio of *E. dalrympleana* to *E. pauciflora* trees is 1:1.12 but in mature stands the ratio changes to

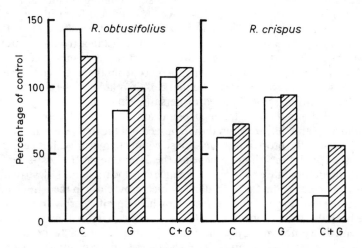

Fig. 6.1 Effects of light grazing by *Gastrophysa viridula* on competition between *Rumex obtusifolius* and *R. crispus*. Open columns represent mean leaf area and shaded columns mean root dry weight. C, competition; G, grazing; C+G, competition and grazing combined. (Reproduced from Bentley and Whittaker [172], by permission of Blackwell Scientific Publications.)

0.38:1. This change during stand development is thought to be brought about by insects and fungi which differentially reduce the growth rate of the two species by damaging the apical shoots. Immature stands of *E. dalrympleana* are subject to much heavier combined attack by insects and fungi than *E. pauciflora*, resulting in a 30% and a 15% effective loss of leaf area respectively. Older stands of both species suffer equivalent losses of about 8% [176].

Insects also feed directly on plant propagules, such as seeds and fruit on which a plant population may ultimately depend for its survival. Sustained high levels of seed predation may potentially alter or regulate the species composition of a plant community. Several insect groups have specialized as 'seed predators' but often their rate of seed removal is too low to have a major impact on the plant community. Nevertheless, even a relatively low rate of seed predation may be important if it is selective. For example, the harvester ant *Veromessor pergandei*, feeds on the seeds of desert ephemerals in the western USA. Total seed removal is less than 8% of total seed production. However, seeds of the dominant plant species *Plantago* suffer proportionately less predation over long time periods than its competitors and this may be a factor contributing to its success [177].

In contrast, the seeds of tropical forest trees are heavily predated both by highly specialized and host-specific insects and by vertebrates. Janzen has hypothesized that the high tree diversity of tropical forests, characterized by the low density and regular spacing of individual species, is maintained by a combination of insect/vertebrate seed predation, which prevents any one species becoming dominant [178]. The suggested mechanism is illustrated in graphical form in Fig. 6.2. This graph describes the spatial dynamics of seed density (I) and the probability of seed survival (P) as a function of distance from the parent tree. As seeds tend to fall vertically seed density is highest below the parent and declines with increasing distance from the tree, producing a typical seed shadow curve (I). The shape of this curve will be determined by the effectiveness of the seed dispersal mechanisms and the rate of viable seed input. The latter depends on the overall rate of seed production and the rate of predation of seeds on the tree. A second curve (P) describes the probability of a seed surviving to maturity. It is claimed that seed predation is highest beneath the parent tree as the host-specific seed predators have a greater chance of locating and destroying individual seeds. Similarly, seedlings which grow close to the parent will be more easily located and destroyed by host-specific herbivores. Thus, the probability of survival to maturity increases with distance from the parent. If we now multiply the I and P curves together to produce a population recruitment curve (PRC) we can see that recruitment reaches a maximum at a certain minimum distance from the parent tree. If this distance is characteristic for a given tree species it will result in a regular tree spacing pattern. Moreover, the gaps between trees will then be available for colonization by other species.

Recent detailed studies on the distribution of trees in the tropical

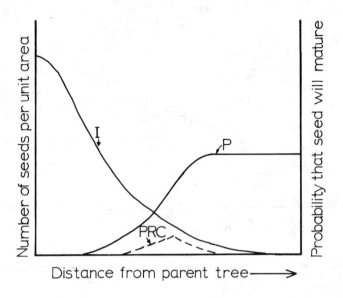

Fig. 6.2 Seed input (I), probability of seed survival (P) and population recruitment curve (PRC) as a function of distance from parent tree in a tropical rain forest. (From Janzen [178], by permission of University of Chicago Press.)

forests of Costa Rica throw serious doubts on the credibility of this elegantly simple hypothesis [179]. All 114 tree species examined showed aggregated or random distributions, none were spaced regularly and only a few species showed a reduced number of juvenile trees close to the adults. The sapling mortality of the 30 commonest species appeared to be a random thinning process and in some species mortality actually increased with distance from parent. These results suggest that, despite heavy seed and seedling predation, a minimum inter-tree distance did not operate, or was less than one tree crown diameter and that some trees reach maturity adjacent to existing adults. Rates of seed production and seed predation appeared to be so variable that no clear relationship between seed predation and tree density could be established. Thus, factors other than a spacing constraint seem to be important in preventing the dominance of single species. Nevertheless, this does not rule out the possibility that seed predators may act as frequency-dependent regulators of tree abundance and thereby help to maintain diversity.

Janzen's original graphical model was not scaled and the failure accurately to describe the Costa Rican forest may be a problem of scaling. Fig. 6.3 shows a rescaled version of the model, applicable to Costa Rica, in which the I and P curves are adjusted so that the resulting PRC values are highest below the parent and decline with increasing distance. There is as a result no characteristic minimum distance between trees of the same species.

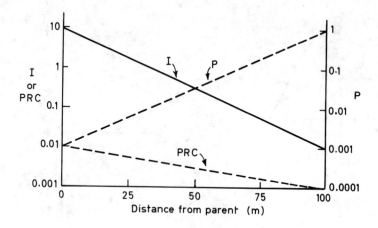

Fig. 6.3 Scaled version of Janzen's model for Costa Rican forest. (Reproduced from Hubbell [179], with permission.)

7 Insect herbivory in ecosystems

Although mentioned, crucial role of enemies is not well taken into consideration.

7.1 The scale of insect herbivory

Whole ecosystem studies are complex and time consuming and it is not feasible to study every organism. Ecologists have usually concentrated on the apparently dominant groups of insect herbivores feeding on the dominant plant species. Often whole insect groups, particularly the sap-sucking fauna, have been ignored. For example, most studies on grassland or rangeland have concentrated on grasshoppers at the expense of other groups [201, 202]. Insects feeding below ground have almost invariably been disregarded. Nevertheless, many data are available which indicate the scale of insect herbivory in a variety of natural ecosystems, ranging from the high arctic tundra to the rain forests of the tropics.

Natural ecosystems differ widely both in diversity and productivity. For example, tundra ecosystems usually exhibit low plant species diversity, low levels of primary production and support few insect herbivore species. Tropical rain forests, by contrast, exhibit high plant diversity, high productivity and support a highly diverse insect fauna. Table 7.1 summarizes data for a range of ecosystems, showing the consumption of above ground primary production by insects, expressed in energy units (kJ m^{-2}). Occasionally it has been necessary to use appropriate tabulated values to convert published biomass figures to energy units [22]. The data indicate the scale of insect consumption in different ecosystems but they are limited with respect to different experimental approaches and methodology and should be interpreted with caution.

Different experimental methods have been used to estimate consumption by the insects. The most direct method involves measuring the amount of leaf tissue removed from the plant by the insects. This is more difficult than might appear as leaf holes grow and it is difficult to equate holes of different ages. Also it is difficult to account for leaves which have been eaten completely [187]. Experimental results for trees indicate that 'leaf hole' methods can overestimate percentage leaf consumption by a factor of 1.5–4. This is a function of both tree species and the biology of the insect consumer. Errors are more likely to occur for tree species which expand their leaves over extended periods as opposed to species which expand their leaves in a single short synchronous burst [187]. The method also fails to account for material removed by sap-sucking insects. An alternative method, which is less direct, is first to estimate the insect population. Feeding rates or energy conversion ratios (Chapter

3), derived from laboratory studies, can then be applied to the population estimate to arrive at a figure for total consumption. A variation of this method is to sample the faeces produced by the insect population and apply a known faeces produced:consumption ratio to calculate consumption.

The portion of the total primary production which is available to the insect varies, dependent on the growth form of the vegetation. For herbaceous plant communities such as grassland the total above-ground production represents the resource available to the insects. In contrast, much of the above-ground production in communities containing woody species is diverted into the woody tissues where it becomes less available to the insects. Leaves, new shoots and reproductive structures represent the 'available' production. Consumption by insects in forest ecosystems has often been estimated as a proportion of the available primary production rather than as a smaller proportion of the total primary production. The ratio of the net production of leaves to the total above ground production varies between tree species: values in Table 7.1 lie between 2.1–4.4. This can present problems when comparing data for different ecosystems.

For the sake of clarity, information in Table 7.1 is presented under a number of separate headings. Available primary production is distinguished from total primary production. The total consumption figure, despite limitations, is based on the best available estimate and percentage consumption is expressed in three separate ways. Estimates of the percentage of the available production which is consumed are separated into those based on leaf-hole methods and those obtained by other less direct methods. The final column is an estimate of the consumption as a percentage of the total primary production. The figures in parentheses represent our educated guess at the correct figure.

What conclusions can be drawn from the table? First there is a significant correlation ($r = 0.78$) between consumption and available production when both are expressed on a logarithmic scale (Fig. 7.1). This indicates that as productivity rises so does the amount consumed. There is little evidence, however, to suggest any strong relationship between ecosystem diversity and percent consumption.

Consumption of available primary production appears to lie within a range of 0–15% with a mean of about 5%. When expressed as a percentage of total production consumption seldom rises above 10%. It has been argued that Australian *Eucalyptus* communities are subject to unusually high levels of insect herbivory with levels of consumption, measured by the leaf-hole method exceeding 30% [203]. However, when the necessary correction factors are applied the figure begins to overlap the upper range for other ecosystems.

7.2 The role of insect herbivores in the ecosystem

Virtually all ecosystems consist of two main trophic or food chain pathways, a herbivore pathway based on the consumption of living

Table 7.1 Comparison of consumption of plants by insects in different ecosystem types. NPP = total net primary production. ANPP = available net primary production. C = consumption. Consumption is also expressed as a percentage of both NPP and ANPP. Estimates of % consumption of ANPP are divided into those obtained by indirect methods[1] and leaf hole methods[2]. An asterisk indicates data to which an energy conversion factor has been applied

Vegetation type	Country	Dominant vegetation	NPP (kJ m^{-2})	ANPP (kJ m^{-2})	C (kJ m^{-2})	%C ANPP [1]	%C ANPP [2]	%C NPP	Reference
Tundra	Canada	Dryas/sedge meadow	1917	1917	1.2–3.5	—	—	—	[180]
Blanket bog	England	Heather/cotton grass	8 482					(>3)	[181]
Coniferous forest	Japan	Mixed conifers	16 794*	5 707*	351*	6	—	2	[182]
Deciduous forest (temperate)	England	Oak	—	7 289*	0–1087*	0–13	—	—	[183]
	England	Oak/hazel	—	5 641*	16–35	0.3–0.9	5–7	—	[184]
	England	Hazel coppice	—	2 935	118–130	4	—	5	[92]
	Finland	Birch	3 244*	1 135*	166*	15	—	1	[185]
	Denmark	Beech	26 400	5 940	95–301	2–5	6–19	—	[134]
	Poland	Pine/Vaccinium	—	6 154	569	—	9	—	[186]
	Poland	Pine/oak	—	7 411	280	—	4	—	[186]
	USA	Tulip tree	9 791	4 181	79–142	2–3	5–10	1–2	[187]
Montane grassland	New Zealand	Grasses	1 004*	1 004	24*	2	—	2	[188]
	Japan	Grasses	3 048	3 048*	18–44*	0.6–1.4	—	—	[189]
Temperate grassland (cool)	Canada	Grasses	2 574	2 574	52	2	—	2	[88]
	Finland	Grasses	8 944	8 944	133–192	1–3	—	1–3	[190]
Temperate grassland (warm)	USA	Grasses	5 359	5 359	471	9	—	9	[89]
	USA	Grasses	10 688	10 688	95	1	—	1	[191]
Salt marsh	USA	Grasses	27 570	27 570	2177	8	—	8	[192]
Old field	USA	Herbs/grasses	5 694	5 694	25	0.4	—	0.4	[193]
	USA	Herbs/grasses	4 500	4 500	356	8	—	8	[194]
Desert	USA	Creosote bush	—	78*	0.7–1.6*	1–2	—		[195]
Sclerophyll forest	Australia	Eucalyptus	—	—	—	—	15–30	—	[196]
	Australia	Eucalyptus	—	—	—	—	8–33	—	[197]
Tropical forest	Panama	Mixed forest	23 472*	14 761*	1269*	—	9	—	[198]
	Puerto Rico	Mixed forest	—	11 014*	705*	—	6	—	[199]
	India	*Shorea* forest	18 635	—	1590	—	—	9	[200]

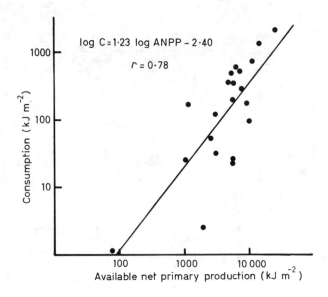

Fig. 7.1 Relationship between consumption by insects and available net primary production based on data in Table 7.1.

plant tissue and a saprovore pathway based on the consumption of dead plant material. There are important differences between the two pathways. Herbivores can interact directly with the plant and modify the rate of energy input into the system: saprovores have no direct effect on primary productivity but may influence it indirectly through their role in nutrient recycling [204].

It is clear from the previous section that the amount of energy passing to the insect consumers usually represents less than 10% of the net primary production. Of this at least half is passed almost immediately to the saprovores in the form of faeces or honeydew. The major energy flux, therefore, is via the saprovore food chain.

This raises the question of the functional role of insect herbivory in the dynamics of ecosystems. Are the herbivorous insects a minor nuisance against which plants have evolved effective defence mechanisms or do they play a more significant role?

The first major potential role of the insect herbivore is in suppressing the productivity of the primary producers. The simple measurement of total energy consumed tells us little about their functional role and their effect on a plant may be much higher than we would predict from a knowledge of their feeding rate. For example, at high population levels the sycamore aphid *Drepanosiphum platanoides* feeding on sycamore (*Acer pseudoplatanus*) causes a reduction in leaf area six times greater than that which would be expected from calculations based on its ingestion rate. This may cause suppression of stem-wood growth by up to 64% [205]. Similarly, if we examine the energy consumed and the drain of essential nutrients similar discrepancies may occur. The aphid *Macrosiphum liriodendri* feeding on *Liriodendron tulipifera*, the tulip tree, removes only 1% of annual photosynthate production but this represents 17% of the annual standing crop of foliar nitrogen [104].

As insects feed, the faeces or honeydew produced represents a premature release of nutrients to the saprovore system, often in a readily available form. After mineralization, the nutrients become reavailable to the plant for growth, possibly within the same growing season. The insect is, therefore, acting to short-circuit the normal nutrient cycling pathway. Caterpillar faeces or frass falling onto the soil surface may contain up to 25% of water soluble components which are readily available to the microflora. In litter bag experiments the addition of frass to forest litter often appears to stimulate the rate of litter decomposition and thereby the rate at which nutrients are released [206]. The honeydew produced by aphids and other Homoptera is rich in sugars and serves as a prime food source for a large array of adult insects [207]. Recent experiments suggest that the sugars in aphid honeydew stimulate the activity of soil microflora particularly the nitrogen fixing forms. If this is correct then aphids may actively increase the availability of inorganic nitrogen to their host-plants [209, 210].

The theory has recently been proposed that insect herbivores may act as regulators of primary production in forest ecosystems, so that over long time periods they ensure a consistent and optimal level of plant production at a given site [211, 212]. Insect herbivores possess several characteristics which could fit them to this role. First, they occupy the strategic trophic position linking the primary producers to the nutrient recycling saprovore system. Secondly, they are closely coevolved and coadapted with their host-plant species and can change their population density and feeding impact either positively or negatively in response to changes in the state of the plant. Finally, as their population density changes they can induce parallel changes in the growth and physiology of their host-plant.

The theory states that the intensity of insect herbivory varies as an inverse function of host plant vigour and productivity. Young, vigorous and productive plants are only marginally adequate as a food source and support only a low level of insect herbivory. Loss of vigour occurs in older less productive plants and those growing in unfavourable environments subject to stresses such as nutrient depletion or water stress. However, these plants which in a sense are less able to defend themselves are a better food source for the insects. This enhanced food supply will, in turn, lead to an increase in the insect population which the plant can support and a higher feeding pressure. Ultimately, this results in a greater input of nutrients to the decomposer system and an increase in ecosystem productivity.

Why do changes in plant palatability occur? There is much evidence to suggest that when plants are subjected to environmental stress, there is an increase in the soluble nitrogen levels and a decrease in the concentrations of secondary compounds of their leaf tissues. This may produce a direct increase in their palatability to insects [213, 214, 215]. The 'total quantity' of available food is, therefore, a function of both plant quality and plant biomass [216].

Much of the evidence supporting the regulation theory is based on correlations or on anecdotal information rather than direct experimentation. For example, in temperate forest ecosystems, which normally suffer low level defoliation, outbreaks of insects usually affect older or diseased trees and often occur on sites which suffer from nutrient deficiency, water stress or periodic waterlogging [212]. *Eucalyptus* forests in Australia which appear to sustain higher chronic levels of defoliation, grow on infertile soils and it has been argued that the high level of defoliation is an adaptation to maintain soil fertility [196]. Furthermore, there is an apparent correlation between outbreaks of *Eucalyptus* feeding psyllids and the potential water stress on the plant, calculated from meteorological data (Fig. 7.2) [213]. This same argument, however, does not appear to hold for tropical rain forest ecosystems which grow under similar nutrient-limited conditions and which, during an irregular dry season might be expected to suffer from water stress. These ecosystems appear to incur only moderate losses to insects.

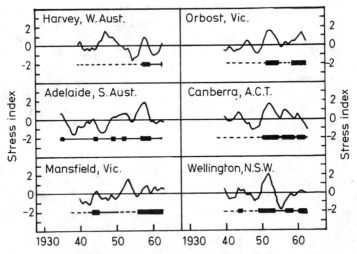

Fig. 7.2 Correlation between water stress index and outbreaks of various species of spondyliaspid psyllids at various sites in Australia. For each site the upper part of the diagram shows changes in the stress index. Positive values indicate periods of water stress on the host plant *Eucalyptus*. The lower part of each diagram represents the historical record of psyllid population density. Outbreak periods are represented by the thick portion of the line. The thin solid lines indicate low population densities and the broken line indicates an absence of records. (Reproduced from White [213], *Ecology* by permission of Ecological Society of America and Duke University Press.)

The theory may also have relevance for ecosystems other than forests. Outbreaks of grass-feeding locust populations often seem to follow irregular periods of water stress or seasonal flooding of the habitat. It has been suggested that they are initiated in areas supporting plant growth which is suppressed but nutritionally superior for locusts [215].

Direct experimental evidence to test the relationship between environment, plant quality, primary production and insect herbivore density is only just beginning to accumulate. A study of the Cinnabar moth (*Tyria jacobaeae*) and its larval host-plant, ragwort (*Senecio jacobaea*), in North America, showed that larval survival and adult fecundity increased with increasing nitrogen content in the food [217]. Data from nine natural populations suggested that changes in host-plant quality were a major factor contributing to fluctuations in moth density. In this case, the natural variations in host-plant quality were thought to be determined by soil moisture characteristics. Detailed studies of insects on trees such as hawthorn, beech and ash, growing along motorway verges in Britain have revealed a correlation between insect population outbreaks and plant nitrogen levels [218]. The mechanisms were thought to involve increased exposure of plants to nitrogen oxides in car exhausts. The effect could be one of direct fertilization of the trees or a more subtle one involving pollutant stress.

Experiments on a red mangrove (*Rhizophora mangle*) ecosystem showed that nutrient enrichment led to higher productivity, increased foliar nitrogen levels, greater insect herbivore population densities and a higher percentage rate of defoliation, the opposite of what the regulation theory would predict [219]. Further contradictory evidence is provided by an experiment on the tent caterpillar *Malacasoma californicum* and its host wild rose (*Rosa nutkana*) [220]. When the plant was subjected to water stress by severing part of the root system there was no measurable change in leaf total nitrogen levels or in larval survival. Soluble nitrogen levels were not measured. *Malacasoma*, however, is probably not the ideal insect on which to test these ideas as it is better adapted for feeding on low quality, mature foliage and seems unlikely to show a significant response to changes in host quality.

The interaction between the moth *Oporinia autumnata* and its host-plant *Betula* indicates the presence of a complex series of defensive reactions by the plant. These reactions, which operate over varying time scales are largely intrinsic to the plant/herbivore association and act independently of climate and soil nutrient status. Increased feeding pressure actually stimulates a defensive response on the part of the plant and it is the plant which appears to regulate the productivity of its insect herbivore rather than vice versa [208].

In conclusion we can say that there is growing evidence to support the idea that insect herbivore populations respond to changes in host-plant quality and that these changes may be moderated by factors both intrinsic and extrinsic to the ecosystem. The examples discussed appear to show that the insect herbivores, rather than acting specifically as regulators of production, are responding directly to changes in host-plant quality and availability. The response probably occurs irrespective of the mechanism which brings the change about, whether it is climatic stress, nutrient stress, nutrient enhancement or a more subtle intrinsic relaxation of the plant's anti-herbivore defences. Nevertheless, where such relationships exist insect herbivores may, in some ecosystems, act as crude regulators of primary production.

It should be remembered, of course, that not all insects may respond in the manner described. Here we have looked specifically at insect–plant relationships and have largely ignored the other factors which may influence the population dynamics of insects, such as natural enemies and the direct effect of climate [221]. At the ecosystem level of organization we are dealing with extremely complex systems of which the insect–plant relationship is only one important component. The extent to which the diversity and complexity of ecosystem determines the observed response remains unclear. Why for instance, is the rate of herbivory low in some nutrient limited, waterlogged tundra ecosystems? Do herbivorous insects really regulate production, is the food just too unpalatable or is the climate too severe to permit sustained periods of insect population growth? In tropical rain forests is the response of an individual insect–plant association masked by the high background diversity of the ecosystem?

7.3 Insect herbivory and agricultural ecosystems

It has been estimated that an amount equivalent to about 21% of current world food production is lost to plant-feeding insects. In developing countries where fewer control measures are applied, percentage losses may be much higher [222]. Harvested agricultural crops usually consist of only a selected proportion of the total primary production, such as leaves, seeds or storage organs. Nevertheless, the general level of insect herbivory in the absence of control measures is much higher in man-made agricultural ecosystems than in natural ecosystems. An understanding of the ways in which insect–plant relationships differ between such ecosystems is, therefore, of fundamental importance in formulating control strategies for insect pests.

Pest species are often those naturally occurring species which by reason of their biology are preadapted to exploit new man-made ecosystems. For example, in its natural habitat in the western USA the Colorado beetle *Leptinotarsa decemlineata* feeds at relatively low density on wild members of the plant family Solanaceae. The transition to pest status occurred only when man began planting on a large scale a highly acceptable food source, potato, thereby creating large areas of favourable habitat.

Agricultural crops, can be classified as non-apparent species (as defined in Chapter 2) growing at high densities. Such plants are usually eaten by polyphagous insect species, which in natural ecosystems are adapted for living on ephemeral and less predictable food resources.

These insects often have a very high reproductive rate which offsets the high mortalities incurred in locating their food plants. Furthermore, the natural host-plants can act as food refuges, supporting the insect population during periods when a suitable crop is unavailable.

The major differences between natural and agricultural ecosystems include both differences in the individual plants themselves and the way the plants are arranged within the ecosystem. As we have noted previously, the level of insect herbivory on a particular plant species is affected by deterrent chemicals and physical defences and by the nutrient status of the plant, particularly the nitrogen levels. Often these plant properties have been altered during the domestication and development of agricultural varieties. The secondary compounds and tough structural materials which deter insect feeding often reduce the acceptability and palatability of the plant to human beings and, therefore, improved varieties have been produced with low levels of these compounds. Furthermore, attempts to improve the nutritional value of plants often aim to raise the protein nitrogen levels of the tissues, making them more acceptable to insect consumers. Such improvements have been permitted by the use of insecticides to reduce the selection pressure of the insect on the plant. Many modern plant varieties have been bred to give high yield in soils treated with fertilizers. Wher these varieties are grown in suboptimal conditions they may be more susceptible to insect attack.

Most agricultural crops are grown as a monoculture, that is a single species plant community with an even age structure, grown at high density. In such low-diversity systems the chance of locating a host-plant and subsequent reproduction by a given individual insect is greatly increased. If the insect is a plant disease vector then its effectiveness in disease transmission is similarly enhanced and its overall impact on the ecosystem will be increased. Structurally simple monocultures tend to support a lower diversity and density of insect predators than natural ecosystems, with the result that predation on the insect herbivore population will be reduced.

Many plant species contribute to the total primary production of diverse natural ecosystems and each species has its own group of associated insect herbivores. If, therefore, one particular plant species were to suffer heavy losses to insects the effect on the primary production of the ecosystem would be small. In contrast, in a monoculture, there would be drastic effects on overall productivity. Thus the diversity and complexity of the food web structure can act as a buffer which contributes stability to the overall productivity of the ecosystem.

If, in the future, we are to become less dependent on the use of insecticides for the control of agricultural pests the above considerations must be taken into account in the design and management of agricultural ecosystems.

References

[1] Tenow, O. (1972), *Zool. Bidr. Upps.*, Suppl. 2, 1–107.

[2] Wolda, H. and Foster, R. (1978), *Geo-Eco-Trop.*, **2**, 443–454.

[3] Malyshev, S.I. (1968), *Genesis of the Hymenoptera and the phases of their evolution*, Methuen, London.

[4] Smart, J. and Hughes, N.F. (1973), *Symp. R. Ent. Soc. Lond.* **6**, 143–155.

[5] Slansky, F. (1976), *J. N.Y. Ent. Soc.*, **84**, 91–105.

[6] Rowell, H.F. (1978), *Entomologia Exp. Appl.*, **24**, 651–662.

[7] Pitkin, B.R. (1976), *Ecol. Entomol.*, **1**, 41–47.

[8] Eastop, V.F. (1973), *Symp. R. Ent. Soc. Lond.*, **6**, 157–178.

[9] Janzen, D.H. (1980), *J. Ecol.*, **68**, 929–952.

[10] Claridge, M.F. and Wilson, M.R. (1981), *Ecol. Entomol.*, **6**, 217–238.

[11] Hodkinson, I.D. and White, I.M. (1979), *Handbk. Ident. Br. Insects*, **2**(5a), 1–98.

[12] Benson, R.B. (1951–58), *Handbk. Ident. Br. Insects*, **2**(a), 1–49; **2**(b), 1–137; **2**(c), 1–252.

[13] Mulkern, G.B. (1967), *A. Rev. Ent.* **12**, 59–78.

[14] Powell, J.A. (1980), *A. Rev. Ent.* **25**, 133–159.

[15] Holloway, J.D. and Herbert, P.D.N. (1979), *Biol. J. Linn. Soc.*, **11**, 229–251.

[16] Jolivet, P. and Petitpierre, E. (1976), *Annls. Soc. Ent. Fr.* (N.S.), **12**, 123–149.

[17] Myerscough, P.J. (1980), *J. Ecol.*, **68**, 1047–1074.

[18] Claridge, M.F. and Wilson, M.R. (1978), *Am. Nat.*, **112**, 451–456.

[19] Southwood, T.R.E. (1973), *Symp. R. Ent. Soc. Lond.*, **6**, 3–30.

[20] Hughes, M.K. (1971), *Ecology*, **52**, 923–926.

[21] Gorham, E. and Sanger, J. (1967), *Ecology*, **48**, 492–494.

[22] Cummins, K.W. and Wuycheck, J.C. (1971), *Mitt. Internat. Veirin Limnol.*, **18**, 1–158.

[23] Allen, S.E., Grimshaw, H.M., Parkinson, J.A. and Quarmby, C. (1974), *Chemical Analysis of Ecological Materials*, Blackwell, Oxford.

[24] Mattson, W.J. (1980), *A. Rev. Ecol. Syst.*, **11**, 119–161.

[25] Hughes, M.K., Lepp, N.W. and Phipps, D.A. (1980), *Adv. Ecol. Res.*, **11**, 217–327.

[26] Morrison, F.B. (1949), *Feeds and feeding*, Morrison Publishing Company, Ithaca.

[27] Haukioja, E., Niemela, P., Iso-Iivari, L., Ojala, H. and Aro, E.M. (1978), *Rep. Kevo Subarctic Res. Stn.*, **14**, 5—12.

[28] McNeill, S. (1973), *J. Anim. Ecol.* **42**, 495–507.

[29] McNeill, S. and Southwood, T.R.E. (1978), in *Biochemical Aspects of Plant and Animal Coevolution* (Harborne, J.B., ed.), Academic Press, London, pp. 77–98.

[30] Parry, W.H. (1976), *Oecologia (Berl.)*, **23**, 297–313.

[31] Mittler, T.E. (1958), *J. Exp. Biol.,* **35**, 74–84.

[32] Woodwell, G.M., Whittaker, R.H. and Houghton, R.A. (1975), *Ecology,* **56**, 318–322.

[33] Satchell, J.E. (1962), *Ann. Appl. Biol.,* **50**, 431–442.

[34] Feeney, P.P. (1970), *Ecology,* **51**, 565–581.

[35] Levin, D.A. (1973), *Q. Rev. Biol.,* **48**, 3–15.

[36] Pillemer, E.A. and Tingey, W.M. (1978), *Entomologia Exp. Appl.,* **24**, 83–94.

[37] Janzen, D.H. (1966), *Evolution,* **20**, 249–275.

[38] Janzen, D.H. (1967), *Kans. Univ. Sci. Bull.,* **67**, 315–558.

[39] Inouye, D.W. and Taylor, O.R. (1979), *Ecology,* **60**, 1–7.

[40] Robinson, T. (1979), in *Herbivores. Their interaction with secondary plant compounds* (Rosenthal, G.A. and Janzen, D.H., eds), Academic Press, New York, pp. 413–448.

[41] Dolinger, P.M., Ehrlich, P., Fitch, W. and Breedlove, D. (1973), *Oecologia* (Berl.), **13**, 191–204.

[42] Gustafsson, A. and Gadd, I. (1965), *Hereditas,* **53**, 15–39.

[43] Harley, K.L.S. and Thorsteinson, A.J. (1967), *Can. J. Zool.,* **45**, 305–319.

[44] Kuhn, R. and Loew, I. (1947), *Chem. Ber.,* **80**, 406–410.

[45] Kuhn, R. and Loew, I. (1961), *Chem. Ber.,* **94**, 1088–1095.

[46] Rhoades, D.F. (1979), in *Herbivores. Their interaction with secondary plant metabolites* (Rosenthal, G.A. and Janzen, D.H., eds), Academic Press, New York, pp. 3–54.

[47] van Etten, C.H. and Tookey, H.L. (1979), in *Herbivores. Their interaction with secondary plant metabolites* (Rosenthal, G.A. and Janzen, D.H., eds), Academic Press, New York, pp. 471–500.

[48] Joseffson, E. (1967), *Phytochemistry,* **6**, 1617–1627.

[49] Conn, E.E. (1979), in *Herbivores. Their interaction with secondary plant metabolites* (Rosenthal, G.A. and Janzen, D.H., eds), Academic Press, New York, pp. 387–412.

[50] Stama, K. (1979), in *Herbivores. Their interaction with secondary plant metabolites* (Rosenthal, G.A. and Janzen, D.H., eds), Academic Press, New York, pp. 683–700.

[51] Feeney, P.P. (1968), *J. Insect Physiol.,* **14**, 805–817.

[52] Feeney, P.P. (1976), *Recent Adv. Phytochem.,* **10**, 1–40.

[53] Rhoades, D.F. and Cates, R.G. (1976), *Recent Adv. Phytochem.,* **10**, 168–213.

[54] Swain, T. (1979), in *Herbivores. Their interaction with secondary plant metabolites* (Rosenthal, G.A. and Janzen, D.H., eds), Academic Press, New York, pp. 657–682.

[55] Bernays, E.A. (1981), *Ecol. Entomol.,* **6**, 353–360.

[56] Bernays, E.A. (1978), *Entomologia Exp. Appl.,* **24**, 244–253.

[57] Fox, L.R. and Macauley, B.J. (1977), *Oecologia* (Berl.), **29**, 145–162.

[58] Benz, G. (1977), *Eucarpia/IOBC Wking Group Breed Resistance Insects Mites Bull SROP,* 1977/8 Report, pp. 155–159.

[59] Baltensweiler, W., Benz, G. Bovey, P. and Delucchi, V. (1977), *Annu. Rev. Ent.,* **22**, 79–100.

[60] Haukioja, E. and Niemela, P. (1976), *Rep. Kevo Subarctic Res. Stn,* **13**, 44–47.

[61] Feeney, P.P. (1975), in *Coevolution of Animals and Plants,* (Gilbert, L.E. and Raven, P.H., eds), University of Texas Press, Austin, pp. 1–19.

[62] Owen, D.F. and Wiegert, R.G. (1981), *Oikos*, **36**, 376–378.
[63] Caldwell, M.M., Richards, J.H., Johnson, D.A., Nowak, R.S. and Dzurec, R.S. (1981), *Oecologia* (Berl.), **50**, 14–24.
[64] Johnson, C.G. (1969), *Migration and dispersal of insects by flight.* Methuen, London.
[65] Kennedy, J.S. (1976), in *The host plant in relation to insect behaviour and reproduction* (Jermy, T., ed.), Plenum, New York, pp. 121–123.
[66] Wallbank, B.E. and Wheatley, G.A. (1979), *Ann. Appl. Biol.*, **91**, 1–12.
[67] Dethier, V.G. (1976), in *The host plant in relation to insect behaviour and reproduction* (Jermy, T., ed.), Plenum, New York, pp. 67–70.
[68] Saxena, K.N. and Saxena, R.C. (1975), *Entomologia Exp. Appl.*, **18**, 207–212.
[69] Tanton, M.T. (1977), *Entomologia Exp. Appl.*, **22**, 113–122.
[70] Cook, A.G. (1977), *Ecol. Entomol.*, **2**, 113–121.
[71] Bernays, E.A. and Chapman, R.F. (1977), *Ecol. Entomol.*, **2**, 1–18.
[72] Chapman, R.F. (1974), *Bull. Ent. Res.*, **64**, 339–363.
[73] Schoonhoven, L.M. (1968), *Annu. Rev. Ent.*, **13**, 115–136.
[74] Hodkinson, I.D., Jensen, T.S. and MacLean, S.F. (1979), *Ecol. Entomol.*, **4**, 119–132.
[75] Wolda, H. (1978), *J. Anim. Ecol.*, **47**, 369–382.
[76] Chapman, R.F. (1969), *The insects: structure and function*, English Univ. Press, London.
[77] Chapman, R.F. (1974), *Feeding in Leaf-Eating Insects*, Oxford Univ. Press.
[78] Pollard, D.G. (1973), *Bull. Ent. Res.*, **62**, 631–714.
[79] Forbes, A.R. and Raine, J. (1973), *Can. Ent.*, **105**, 559–567.
[80] Pollard, D.G. (1972), *J. Nat. Hist.*, **6**, 261–271.
[81] Horsfield, D. (1978), *Entomologia Exp. Appl.*, **24**, 95–99.
[82] Miles, P.W. (1972), *Adv. Insect Physiol.*, **9**, 184–255.
[83] Miles, P.W. (1978), *Entomologia Exp. Appl.*, **24**, 534–539.
[84] David, W.A.L. and Gardiner, B.O.C. (1962), *Bull. Ent. Res.*, **53**, 417–436.
[85] Petrusewicz, K. and Macfadyen, A. (1970), *Productivity of Terrestrial Animals*, Blackwell, Oxford.
[86] Gyllenberg, G. (1969), *Acta. Zool. Fenn.*, **123**, 1–74.
[87] Chlodny, J. (1969), *Ekol. Pol.* (ser. A.), **17**, 391–407.
[88] Bailey, C.G. and Riegert, P.W. (1973), *Can. J. Zool.*, **51**, 91–100.
[89] van Hook, R.J. (1971), *Ecol. Monogr.*, **41**, 1–26.
[90] Gere, G. (1956), *Opusc. Zool., Bpest*, **1**, 29–32.
[91] Schroeder, L. (1973), *Oikos*, **24**, 278–281.
[92] Smith, P.H. (1972), *J. Anim. Ecol.*, **41**, 567–587.
[93] Haukioja, E. and Niemela, P. (1974), *Ann. Zool. Fennici*, **11**, 207–211.
[94] Pruscha, H. (1973), *Sber. Akad. Wiss. Wien, Math–Nat*, **181**, 1–49.
[95] Larsson, S. and Tenow, O. (1979), *Oecologia* (Berl.), **43**, 157–172.
[96] Stockner, J.G. (1971), *J. Fish. Res. Bd. Can.*, **28** 73–94.
[97] Chlodny, J., Gromadzka, J. and Trojan, P. (1967), *Bull. Acad. Pol. Sci. Cl. II Ser. Sci. Biol.*, **15**, 743–747.
[98] Axelsson, B., Bosatta, E., Lohm, U., Persson, T. and Tenow, O. (1974), *Zoon*, **2**, 49–55.
[99] Nakamura, M. (1965), *Jap. J. Ecol.*, **15**, 1–18.
[100] Wightman, J.A. (1978), *J. Anim. Ecol.*, **47**, 117–129.
[101] Andrzejewska, L. (1967), in *Secondary productivity of terrestrial eco-*

systems (Petrusewicz, K., ed.), Polish Academy of Sciences, Warsaw, pp. 791–806.

[102] Hinton, J.M. (1971), *Oikos,* **22**, 155–171.

[103] Hodkinson, I.D. (1973), *J. Anim. Ecol.,* **42**, 565–583.

[104] van Hook, R.I., Nielsen, M.G. and Shugart, H.H. (1980), *Ecology*, **61**, 960–975.

[105] Axelsson, B., Lohm, U., Nilsson, A., Persson, T. and Tenow, O. (1975), *Zoon*, **3**, 71–84.

[106] Llewellyn, M. and Qureshi, A.L. (1978), *Entomologia Exp. Appl.,* **23**, 26–39.

[107] Scriber, J.M. and Slansky, F. (1981), *Annu. Rev. Ent.,* **26**, 183–212.

[108] Slansky, F. and Feeny, P. (1977), *Ecol. Monogr.,* **47**, 209–228.

[109] Scriber, J.M. (1977), *Oecologia* (Berl.), **28**, 269–287.

[110] Way, M.J. and Cammell, M. (1970), in *Animal populations in relation to their food resources* (Watson, A., ed.), Blackwell, Oxford, pp. 229–247.

[111] White, T.C.R. (1970), *Aust. J. Zool.,* **18**, 273–293.

[112] Hargreaves, C.E.M. and Llewellyn, M. (1978), *J. Anim. Ecol.,* **47**, 605–614.

[113] Wratten, S.D. (1978), *Ann. Appl. Biol.,* **90**, 11–20.

[114] Chiang, H.C. (1964), *Entomologia Exp. Appl.,* **7**, 144–178.

[115] Bardner, R. (1968), *Ann. Appl. Biol.,* **61**, 1–11.

[116] Jones, F.G.W., Dunning, R.A. and Humphries, K.P. (1955), *Ann. Appl. Biol.,* **43**, 63–70.

[117] Rautappää, J. (1970), *Soumi, Hyont. Aikak*, **36**, 145–152.

[118] Johnson, C.G., Lofty, J.R. and Cross, D.J. (1969), *Rep. Rothamsted Exp. Stn.* (1968), Part 2, 141–156.

[119] van Emden, H.F. (1966), *Entomologia Exp. Appl.,* **9**, 444–460.

[120] Ridsdill Smith, T.J. (1977), *J. Appl. Ecol.,* **14**, 73–80.

[121] Cromatie, W.J. (1975), *J. Appl. Ecol.,* **12**, 517–535.

[122] Pimentel, D. (1961), *Ann. Ent. Soc. Am.,* **54**, 61–69.

[123] Mayse, M.A. (1978), *J. Appl. Ecol.,* **15**, 439–450.

[124] Rausher, M.D. and Feeny, P. (1980), *Ecology*, **61**, 905–917.

[125] Bardner, R., Fletcher, K.E. and Huston, P. (1970), in *Proceedings of the 5th Insecticide and Fungicide Conference (1969)*, British Crop Protection Council, London, pp. 500–504.

[126] Mitchell, N.D. (1977), *Entomologia Exp. Appl.,* **22**, 208–219.

[127] Windle, P.N. and Franz, E.H. (1979), *Ecology,* **60**, 521–529.

[128] Tammes, P.M.L. (1961), *Tijdschr. Pl. Zeikt.,* **67**, 257–263.

[129] Bardner, R. and Fletcher, K.E. (1974), *Bull. Ent. Res.,* **64**, 141–160.

[130] Rautappää, J. (1966), *Ann. Agric. Fenn.,* **5**, 334–341.

[131] Harris, P. (1974), *Agro-Ecosystems*, **1**, 219–225.

[132] Banks, C.J. and Macauley, E.D.M. (1967), *Ann. Appl. Biol.,* **60**, 445–453.

[133] Nielsen, B.O. and Ejlersen, A. (1977), *Ecol. Entomol.,* **2**, 293–299.

[134] Nielsen, B.O. (1978), *Oikos*, **31**, 273–279.

[135] Llewellyn, M. (1975), *J. Appl. Ecol.,* **12**, 15–23.

[136] Kulman, H.M. (1971), *Annu. Rev. Ent.,* **16**, 289–324.

[137] Wickman, B.E., Mason, R.R. and Thompson, C.G. (1973), *USDA Forest Service General Technical Report, PNW-5*, 1–18.

[138] Koslowski, T.T. (1969), *J. For.* **67**, 118–122.

[139] Dixon, A.F.G. (1973), *Biology of Aphids*, Edward Arnold, London.

[140] Wallace, P.P. (1945), *Bull. Conn. Agric. Exp. Stn*, **488**, 358–373.

[141] Heichel, G.H. and Turner, N.C. (1976), in *Perspectives in Forest Entomology* (Anderson, J.F. and Kaya, H.K., eds), Academic Press, New York, pp. 31–40.

[142] Ericsson, A., Larsson, S. and Tenow, O. (1980), *J. Appl. Ecol.,* **17**, 747–769.

[143] Furuno, T. (1975), *Bulletin of Kyoto University Forests*, **47**, 1–14 (Japanese, Eng. sum.)

[144] Rook, D.A. and Whyte, A.G.D. (1976), *N.Z. J. For. Sci.,* **6**, 40–56.

[145] Ericsson, A. (1979), *Physiologia Pl.,* **45**, 270–280.

[146] Waloff, R. and Richards, O.W. (1977), *J. Appl. Ecol.,* **14**, 787–798.

[147] Hughes, M.K. (1971), *Oikos*, **22**, 62–73.

[148] Morrow, P.A. and LaMarche, V.C. (1978), *Science*, **201**, 1244–1246.

[149] Pryor, L.D. (1976), *Biology of Eucalypts*, Edward Arnold, London.

[150] Embree, D.G. (1967), *Forest Sci.,* **13**, 295–299.

[151] Varley, G.C. (1978), in *Dendrochronology in Europe* (Fletcher, J., ed.), BAR International Series 51, pp. 179–182.

[152] Varley, G.C. and Gradwell, G.R. (1962), *Proc. 11th Int. Congr. Ent. (Vienna)*, **2**, 211–214.

[153] Puritch, G.S. and Talmon-de l'Armee, M. (1971), *Can. J. Bot.,* **49**, 1219–1223.

[154] Puritch, G.S. and Johnson, R.P.C. (1971), *J. Exp. Bot.,* **22**, 953–958.

[155] Puritch, G.S. and Petty, J.A. (1971), *J. Exp. Bot.,* **22**, 946–952.

[156] Puritch, G.S. (1971), *J. Exp. Bot.,* **22**, 936–945.

[157] Redmond, D.R. (1959), *Forest Sci.,* **5**, 64–69.

[158] Belyea, R.M. (1952), *J. For.,* **50**, 729–738.

[159] Churchill, G.B., John, H.H., Duncan, D.P. and Hodson, A.C. (1964), *Ecology*, **45**, 630–633.

[160] Blais, J.R. (1965), *Forest Sci.,* **11**, 130–138.

[161] Readshaw, J.L. and Mazanec, Z. (1969), *Aust. For.,* **33**, 29–36.

[162] Schweingruber, F.H. (1979), *Schweiz. Z. Forstwes.,* **130**, 1071–1093.

[163] Harper, J.L. (1977), *Population Biology of Plants*, Academic Press, London.

[164] Lewis, T. (1970), *Ann. Appl. Biol.,* **65**, 213–220.

[165] Lewis, T. (1969), *J. Appl. Ecol.,* **6**, 443–452.

[166] Bowden, J. and Dean, G.J.W. (1977), *J. Appl. Ecol.,* **14**, 343–354.

[167] Smith, R.W. and Whittaker, J.B. (1980), *J. Anim. Ecol.,* **49**, 537–548.

[168] Tahvanainen, J.O. and Root, R.B. (1972), *Oecologia* (Berl.), **10**, 321–346.

[169] Dempster, J.P. and Coaker, T.H. (1974), in *Biology in pest and disease control* (Price, D. and Solomon, M.E., eds), Blackwell, Oxford, pp. 106–114.

[170] Smith, R.W. and Whittaker, J.B. (1980), *J. Anim. Ecol.,* **49**, 225–236.

[171] Harper, J.L. (1969), *Brookhaven Symp. Biol.,* **22**, 48–62.

[172] Bentley, S. and Whittaker, J.B. (1979), *J. Ecol.,* **67**, 79–90.

[173] Whittaker, J.B. (1979), in *Population Dynamics* (Anderson, R.M., Turner, B.D. and Taylor, L.R., eds), Blackwell, Oxford, pp. 207–222.

[174] Bentley, S., Whittaker, J.B. and Malloch, A.J.C. (1980), *J. Ecol.,* **68**, 671–674.

[175] Carter, W. (1973), *Insects in relation to plant disease*, 2nd edn, Wiley, London.

[176] Burdon, J.J. and Chilvers, G.A. (1974), *Aust. J. Bot.,* **22**, 103–114.

[177] Tevis, L. (1958), *Ecology*, **39**, 695–704.

[178] Janzen, D.H. (1970), *Am. Nat., 104*, 501–528.
[179] Hubbell, S.P. (1980), *Oikos,* **35**, 214–229.
[180] Bliss, L.C. (ed.) (1977), *Truelove Lowland, Devon Island, Canada: A High Arctic Ecosystem,* University of Alberta Press, Edmonton.
[181] Coulson, J.C. and Whittaker, J.B. (1978), in *Ecology of moorland animals* (Heal, O.W. and Perkins, D.F., eds), Ecological Studies, Vol. 27, Springer Verlag, Berlin, pp. 52–93.
[182] Kitazama, Y. (1977), in *Ecosystem analysis of the subalpine coniferous forest of the Shigayama I.B.P. area, central Japan,* JIBP Synthesis, Vol. 15, University of Tokyo Press, pp. 181–188.
[183] Carlisle, A., Brown, A.H.F. and White, E.J. (1966), *J. Ecol.,* **54**, 65–85.
[184] Smith, P.H. unpublished data for Meathop Wood IBP site, England.
[185] Haukioja, E. and Koponen, S. (1975), in *Fennoscandian tundra ecosystems,* Part 2. (Wiegolaski, F.E., ed.), Springer-Verlag, Berlin, pp. 181–188.
[186] Kaczmarek, W. (1967), in *Secondary productivity of terrestrial ecosystems* (Petrusewicz, K., ed.), Polish Academy of Sciences, Warsaw, pp. 663–678.
[187] Reichle, D.E., Goldstein, R.A., van Hook, R.I. and Dodson, G.L.J., (1973), *Ecology,* **54**, 1076–1084.
[188] White, E.G. (1974), *NZ J. Agric. Res.,* **17**, 357–372.
[189] Nakamura, M., Matsumoto, T., Nakamura, K. and Ito, Y. (1975), in *Ecological studies in Japanese Grassland,* JIBP synthesis, Vol. 13, Univ. Tokyo Press, pp. 237–249.
[190] Gyllenberg, G. (1970), *Ann. Zool. Fenn.,* **7**, 283–289.
[191] Menhinick, E.F. (1967), *Ecol. Monogr.,* **37**, 255–272.
[192] Teal, J.M. (1962), *Ecology,* **43**, 614–624.
[193] Wiegert, R.G. and Evans, F.C. (1967), in *Secondary productivity of terrestrial ecosystems* (Petrusewicz, K., ed.), Polish Academy of Sciences, Warsaw, pp. 499–518.
[194] Odum, E.P., Connell, C.E. and Davenport, L.B. (1962), *Ecology,* **43**, 88–96.
[195] Mispagel, M.E. (1978), *Ecology,* **59**, 779–788.
[196] Springett, B.P. (1978), *Aust. J. Ecol.,* **3**, 129–139.
[197] Burdon, J.J. and Chilvers, G.A. (1974), *Aust. J. Bot.,* **22**, 265–269.
[198] Leigh, E.G. and Smythe, N. (1978), in *The ecology of arboreal folivores* (Montgomery, G.G., ed.), Smithsonian Press, Washington, pp. 33–50.
[199] Odum, H.T. and Ruiz-Reyes, J. (1970), in *A tropical rain forest* (Odum, H.T., ed.), Office of Information Service, US Atomic Energy Commission, pp. 169–180.
[200] Misra, R. (1968), *Bull. Int. Soc. Trop. Ecol.,* **9**, 105–118.
[201] Blocker, H.D. (1977), in *The impact of herbivores on arid and semi-arid rangelands,* Proc. 2nd US/Australia Rangeland Panel, Australian Rangeland Society, Perth, pp. 357–376.
[202] Andrzejewska, L. and Gyllenberg, G. (1980), in *Grasslands, systems analysis and man* (Breymeyer, A.I. and Van Dyne, G.M., eds), Cambridge University Press, pp. 201–267.
[203] Morrow, P.A. (1977), in *The role of arthropods in forest ecosystems* (Mattson, W.J., ed.), Springer-Verlag, New York, pp. 19–29.
[204] Heal, O.W. and MacLean, S.F. (1975), in *Unifying concepts in ecology* (van Dobben, W.H. and Lowe-McConnell, R.H., eds), W. Junk, The Hague, pp. 89–108.

[205] Dixon, A.F.G. (1971), *J. Appl. Ecol.*, **8**, 165–179.
[206] Zlotin, R.I. (1971), in *Organismes du sol et production primaire*, IV Colloquium pedobiologiae Dijon, 1970, pp. 455—462.
[207] Zoebelein, G. (1956), *Z. Angew. Ent.*, **38**, 369–416; **39**, 129–167.
[208] Haukioja, E. (1980), *Oikos*, **35**, 202–213.
[209] Owen, D.F. (1980), *Oikos*, **35**, 230–235.
[210] Petelle, M. (1980), *Oikos*, **35**, 127–128.
[211] Chew, R.M. (1974), *Ohio J. Sci.*, **74**, 359–370.
[212] Mattson, W.J. and Addy, N.D. (1975), *Science*, **190**, 515–522.
[213] White, T.C.R. (1969), *Ecology*, **50**, 905–909.
[214] White, T.C.R. (1974), *Oecologia (Berl.)*, **16**, 279–301.
[215] White, T.C.R. (1976), *Oecologia (Berl.)*, **22**, 119–134.
[216] White, T.C.R. (1978), *Oecologia (Berl.)*, **33**, 71–86.
[217] Myers, J.H. and Post, B.J. (1981), *Oecologia (Berl.)*, **48**, 151–156.
[218] Port, G.R. and Thompson, J.R. (1980), *J. Appl. Ecol.*, **17**, 649–656.
[219] Onuf, C.P., Teal, J.M. and Valiela, I. (1977), *Ecology*, **58**, 514–526.
[220] Myers, J.H. (1981), *J. Anim. Ecol.*, **50**, 11–25.
[221] Lawton, J.H. and McNeill, S. (1979), in *Population dynamics* (Anderson, R.M., Turner, B. and Taylor, L.R., eds), Blackwell, Oxford, pp. 223–244.
[222] Cramer, H.H. (1967), *Plant protection and world crop production*, Pflanzenschutz-Nachrichten, Bayer, **20**(1), 1–524.

Index

76